Regional Scale Ecological Risk Assessment

Using the Relative Risk Model

Environmental and Ecological Risk Assessment

Series Editor
Michael C. Newman
College of William and Mary
Virginia Institute of Marine Science
Gloucester Point, Virginia

Published Titles

Coastal and Estuarine Risk Assessment
Edited by
Michael C. Newman, Morris H. Roberts, Jr., and Robert C. Hale

Risk Assessment with Time to Event Models
Edited by
Mark Crane, Michael C. Newman, Peter F. Chapman, and John Fenlon

Species Sensitivity Distributions in Ecotoxicology
Edited by
Leo Posthuma, Glenn W. Suter II, and Theo P. Traas

Regional Scale Ecological Risk Assessment

Using the Relative Risk Model

Edited by
Wayne G. Landis

CRC Press
Taylor & Francis Group
Boca Raton London New York

CRC Press is an imprint of the
Taylor & Francis Group, an **informa** business

CRC Press
Taylor & Francis Group
6000 Broken Sound Parkway NW, Suite 300
Boca Raton, FL 33487-2742

First issued in paperback 2020

ISBN 13: 978-0-367-57824-4 (pbk)
ISBN 13: 978-1-56670-655-1 (hbk)
Library of Congress Card Number 2004051871

This book contains information obtained from authentic and highly regarded sources. Reasonable efforts have been made to publish reliable data and information, but the author and publisher cannot assume responsibility for the validity of all materials or the consequences of their use. The authors and publishers have attempted to trace the copyright holders of all material reproduced in this publication and apologize to copyright holders if permission to publish in this form has not been obtained. If any copyright material has not been acknowledged please write and let us know so we may rectify in any future reprint.

Cover photograph:

The region along the northwest coast of Washington State that forms part of the landscape included in the Cherry Point based risk assessments of the book. The photograph illustrates the variety of land uses and ecotypes in the region, the coastline, and the San Juan Islands in the distance. Photograph by Linda S. Landis.

Library of Congress Cataloging-in-Publication Data

Regional scale ecological risk assessment : using the relative risk model / edited by Wayne G. Landis.
    p. cm. -- (Environmental and ecological risk assessment series)
   Includes bibliographical references (p.).
   ISBN 1-56670-655-6
   1. Ecological risk assessment. I. Landis, Wayne G. II. Series.

QH541.15.R57R44 2004
333.71'4--dc22

                     2004051871

Visit the Taylor & Francis Web site at
http://www.taylorandfrancis.com

and the CRC Press Web site at
http://www.crcpress.com

Preface

My start with ecological risk assessment at the regional scale began with the evaluation of risks within the fjord of Port Valdez, Alaska. I initially thought that the process was going to be very straightforward. The USEPA had produced a framework document and there were published case studies. Although the fjord of Port Valdez was larger than a typical contaminated site, the emphasis in the initial study was on the ballast water treatment plant for the oil tankers. As my research group undertook the study, it quickly became apparent that I was wrong in my initial assessment of the magnitude of the assessment process. As we learned more about the site, it quickly became apparent that there were multiple sources of a variety of chemicals and other stressors that operated on numerous species within the fjord. Howard M. Feder and David G. Shaw of the University of Alaska were instrumental in demonstrating the heterogeneity and complexity of the environment. In order to assess risk within the fjord, a new approach was needed to evaluate risk at a regional scale when multiple stressors and multiple endpoints existing in a heterogeneous landscape was the norm. The relative risk model (RRM) was developed and applied to the fjord.

Nine years later the relative risk model for regional-scale ecological risk assessment has been applied to nine sites in North America, South America, and Australia. The sites include a marine fjord, large and small river watersheds, an urbanized salmon stream, a rain forest, a river and salt pan environment, and a coastal region along the state of Washington. The current state of the art of the RRM includes a Monte Carlo probabilistic analysis of the uncertainties in the ranks, extensive use of geographical information systems, an interactive process with stakeholders and decision makers, and an application of the method to retrospective risk assessment. The patterns of risk can be tested by using site-specific toxicity tests, chemical analyses, or by the analysis of biological community structure.

Other applications of the RRM include the evaluation of the changes in risk patterns following various management options. This application elevates the risk assessment model to an interactive management tool that can be applied at a regional scale.

This book summarizes the development and current state of the art of the use of the relative risk model in the performance of regional-scale ecological risk assessments. The initial chapters present the methodology and the critical nature of the interaction between risk assessors and decision makers. Next, the lead investigators of the studies reviewed in this volume present a current perspective of each program. A chapter on future research needs and applications of the relative risk model concludes the volume.

This collaborative project stands as a progress report on the use of a unique approach to evaluating risk at large spatial scales. Hopefully, this volume will serve as a source book for the methods and applications of regional risk assessment in the management of environmental systems. It will be interesting to see what happens next.

Wayne G. Landis

About the Editor

Since 1989, Wayne G. Landis has been the Director of the Institute of Environmental Toxicology and Chemistry, part of Huxley College of the Environment at Western Washington University. He earned a B.A. in biology in 1974 from Wake Forest University and an M.A. and a Ph.D. in zoology from Indiana University in 1978 and 1979, respectively. Dr. Landis has authored over 100 publications, has made 220 scientific presentations, and has edited three books on environmental toxicology and risk assessment for the American Society for Testing and Materials (ASTM). He has served on numerous committees and consulted for industry, nongovernmental organizations, print and electronic media, and federal (U.S. and Canada), state, provincial, and local governments.

In the 1980s his research at the Chemical Research, Development and Engineering Center at the Aberdeen Proving Grounds included the hydrolysis of organophosphates by enzymes found in protozoa, invertebrates, and the biodegradation of riot control agents for which he received two patents. His contributions in the 1990s while directing the Institute include codevelopment of the Community Conditioning Hypothesis, use of multivariate analysis in microcosm data analysis, creation of the Action at a Distance Hypothesis, and the application of complex systems theory to environmental toxicology. His recent efforts have been to apply ecological risk assessment on regional and landscape scales. This effort has led to the development of the Relative Risk Model for multiple stressor and regional-scale risk assessment. He has been interviewed by the Canadian Broadcasting Company, the British Broadcasting Company, CNN, newspapers, and local stations on a variety of environmental topics.

Dr. Ming-Ho Yu and Dr. Landis have also written the popular textbook *Introduction to Environmental Toxicology,* now in its third edition. Dr. Landis is currently serving on the Board of Editors for *Human and Ecological Risk Assessment* and has been listed as part of the review team for *Environmental Toxicology and Chemistry.* Dr. Landis also served a 5-year term on the Committee on Publications for ASTM, the oversight group for all the journals and books for the society. He is finishing his term as a member of the Board of Directors for the Society of Environmental Toxicology and Chemistry North America and is serving a term as Chair of the Technical Committee. He has been involved in the development of *Integrated Environmental Assessment and Management* and now serves as a member of the Founding Editorial Board.

Acknowledgments

I would like to thank the sponsors for allowing my research group to develop the relative risk model (RRM), our risk assessment research, and its various applications. These sponsors include the Regional Citizens Advisory Council for Prince William Sound, the National Council for Air and Stream Improvement, Washington Department of Natural Resources and Techcominco Trail. The City of Bellingham also provided data and resources for the Squalicum Creek research. I thank Leo Bodensteiner and Tim Hall for providing reviews of the manuscript.

Gene Hoerauf of the Department of Environmental Studies, Huxley College of the Environment, Western Washington University has been an essential component of all the studies depicted in this book since Port Valdez. He has worked tirelessly with our students and others from all over the world to collect the digital data necessary to perform the GIS analysis that has become so necessary. Shawn Boeser also assisted in the later studies.

These studies would not have occurred without the strong initial support of Peter M. Chapman and P. Bruce Duncan. Both were involved in the initial review of the risk assessment that was being conducted for Port Valdez, Alaska. At the time, the relative risk model was completely new and untried. Both of these individuals understood what we were attempting to accomplish and supported our efforts. The first paper (Landis and Wiegers 1997) was a result of Dr. Chapman asking for a short article on the approach for *Human and Ecological Risk Assessment*. That paper has led to a series of other articles for that journal on the use of the RRM. Dr. Duncan was also important in recognizing the importance of risk assessment to the analysis of the Cherry Point region and assisting in the development of this long-term project.

Tim Hall and Carol Piening have been great project officers and colleagues in the course of this program. I really appreciate their support in our pursuit of these studies.

Finally, this undertaking could not have succeeded without the support and patience of Linda and my two daughters, Margaret and Eva.

Wayne G. Landis
Bellingham, Washington

Contributors

Philip Brown
School of Agricultural Science
University of Tasmania
Hobart, Tasmania, Australia

Joy Chen
Jones and Stokes
Bellevue, Washington, U.S.A.

Bruce Duncan
Office of Environmental Assessment
U.S. EPA
Seattle, Washington, U.S.A.

Emily Hart Hayes
Institute of Environmental Toxicology
Huxley College of the Environment
Western Washington University
Bellingham, Washington, U.S.A.

Wayne G. Landis
Institute of Environmental Toxicology
Huxley College of the Environment
Western Washington University
Bellingham, Washington, U.S.A.

Mathew Luxon
Winward Environmental LLC
Seattle, Washington, U.S.A.

April Markiewicz
Institute of Environmental
 Toxicology
Western Washington University
Bellingham, Washington, U.S.A.

Sverker Molander
Environmental Systems Analysis
Chalmers University of Technology
Gothenburg, Sweden

Rosana Moraes
Golder Associates Brasil Ltda
Rio de Janeiro, Brazil

Angela M. Obery
Oregon Department of Environmental
 Quality
Eugene, Oregon, U.S.A.

Jill F. Thomas
National Council for Air and Stream
 Improvement
Anacortes, Washington, U.S.A.

Rachel Walker
School of Agricultural Science
University of Tasmania
Hobart, Tasmania, Australia

Janice Wiegers
Alaska Department of Environmental
 Conservation
Fairbanks, Alaska, U.S.A.

Contents

CHAPTER 1

Introduction

Wayne G. Landis

CONTENTS

Since 1997 the relative risk model (RRM) proposed by Wiegers and Landis (Landis and Wiegers 1997; Wiegers et al. 1998) has been used at seven sites to generate regional risk hypotheses on a variety of scales. These scales have ranged from an urban watershed a few square kilometers in size, to a Brazilian rain forest and large coastal marine areas. The studies incorporate multiple sources of multiple stressors with a variety of endpoints in a diverse landscape that exhibits spatial and temporal heterogeneity.

The purpose of this book is to describe the development and current state of the art in using methods based upon the relative risk model (RRM) for regional ecological risk assessment. This chapter introduces and defines regional-scale ecological risk assessment, presents our baseline model for understanding the dynamics of ecological systems, and introduces the organization of the volume.

1-56670-655-6/05/$0.00+$1.50

REGIONAL RISK ASSESSMENT DEFINED

Ecological risk assessment calculates the probability of an impact to a specified set of assessment endpoints over a defined period of time. In the risk assessment of chemicals, exposure and effects are estimated and the probability of the intersection of those functions is calculated. Impacts typically considered are mortality, chronic physiological impacts, and reproductive effects. Most often these risk assessments deal with single chemicals in such classic cases as pesticides, herbicides, organic solvents, metals, polychlorinated biphenyls, and dioxins. Most often the risk assessments dealt with only one or a few biological endpoints. During the 1990s there was an effort to expand ecological risk assessment to more accurately reflect the reality of the structure, function, and scale of ecological structures. Hunsaker, O'Neill, Suter and colleagues (Hunsaker et al. 1990; Suter 1990; O'Neill et al. 1997) formulated the idea of performing regional risk assessments at a landscape scale.

We (Landis and Wiegers 1997; Wiegers et al. 1998) adopted a definition that naturally incorporates multiple stressors, historical events, spatial structure, and multiple endpoints. Our working definition of a regional-scale risk assessment is: A risk assessment deals at a spatial scale that contains multiple habitats with multiple sources of multiple stressors affecting multiple endpoints, and the characteristics of the landscape affect the risk estimate. Although there may only be one stressor of concern to the decision maker, at a regional scale the other stressors acting upon the assessment endpoints are to be considered.

There have been attempts to perform risk assessments based upon the classical USEPA paradigm (Cook et al. 1999; Cormier et al. 2000), a risk assessment framework originally designed for single chemicals and receptors. A principal difficulty is the incorporation of the spatial structure of the environment; the inherent presence of multiple stressors and the values of stakeholders suggest a variety of important ecological services.

Our development of the RRM was due to our discovery that the classical framework as it existed in 1996 was not applicable to the situation at the fjord of Port Valdez, Alaska. In this instance we had a variety of chemical, physical, and biological stressors within a diverse environment with multiple stakeholder-determined endpoints. Our attempts to construct a conceptual model soon led to enormous complication without an integration that addressed the stakeholder-derived endpoints. This situation resulted in the derivation of the alternative method that is the subject of this book. Fortunately the data collected about Port Valdez, especially the chemical analyses of David Shaw and the understanding of benthic community structure provided by Howard Feder, allowed us the setting in which to develop a robust method. Our other research also prepared us for an alternative way of considering how stressors and ecological systems interact.

A DIFFERENT APPROACH TO UNDERSTANDING ECOLOGICAL SYSTEMS

We were also working under an additional set of design considerations. By the late 1990s, our research group had demonstrated or accepted a number of alternative approaches to understanding ecological systems and the effects of chemical and other stressors upon them. Our collaboration with Robin Matthews and Geoff Matthews in the use of multivariate analysis of microcosm data sets also led to the discovery of the persistence of effects even in simple microcosm systems and led to the formulation of the community conditioning hypothesis. Derived from this is the understanding that ecological systems are complex and historical structures. Both ideas were clearly part of my laboratory's culture at that time. The patch dynamics modeling efforts also led us to an appreciation of the importance of spatial structure in understanding the patterns of effects that a stressor can produce. Each idea is summarized below and a more detailed description can be found in Landis and Yu (2004) and the papers cited in that volume.

THE PROPERTIES OF ECOLOGICAL SYSTEMS

The foundation of the community conditioning hypothesis is that ecological communities retain information about events in their history. Community conditioning is not a theoretical construct but was derived from the multivariate data analysis of a series of microcosm experiments performed as a collaboration with Robin Matthews, Geoff Matthews, and my research group. Information was retained by the treated systems for as long as we could perform the experiments. As currently envisioned, the information can be contained in a variety of formats, from the relative frequencies of alleles in the nuclear or mitochondrial DNA to the dynamics of predatory–prey and competitive interactions. Recovery, defined as return to a previous or control (reference) state, is an illusion of perception or a lack of sufficiently detailed analysis. The many interactions and nonlinear relationships within an ecosystem mean that the history of past events is written into the structure and dynamics. The nonlinear dynamics, time delays, and multiple interactions confer upon ecological systems the property of complexity.

Complex, nonlinear structures have specific properties, listed by Çambel (1993). A few properties particularly critical to how ecosystems react to contaminants are:

1. Complex structures are neither completely deterministic or stochastic, but exhibit both characteristics.
2. The causes and effects of the events the system experiences are not proportional.
3. The different parts of complex systems are linked and affect one another in a synergistic manner.
4. Complex systems undergo irreversible processes.
5. Complex systems are dynamic and not in equilibrium; they are constantly moving targets.

Ecological systems have spatial structure. Research by Spromberg, Johns, and Landis (1998) (Landis and McLaughlin, 2003, 2001, 2000; McLaughlin and Landis, 2000) and later collaboration with J. McLaughlin have used three patch metapopulation models to explore the effects of stressors in spatially structured environments. Three important sets of findings were made.

The first finding is that populations in patches removed from the contamination were affected by the presence of the toxicant. In the case of the linear, persistent toxicant model, the effects were the reduction of the population below carrying capacity and fluctuation in population size.

The second finding is that the simulations incorporating toxicant degradation had several possible discrete outcomes occur from the same set of initial conditions. The range and types of outcome depend on the specifics of toxicant concentration, initial population size, and distance between patches.

Third, it is apparent that not just the chemical concentration, but also the population sizes in the connected patches determine the probabilities of outcomes. Small differences in initial population sizes can drastically alter these probabilities, and these are very simple ecological models.

In conclusion, metapopulation dynamics have several important implications for predicting the impact of chemical toxicants:

1. Effects can be promulgated between patches, even if the toxicant is not transferred. There is action at a distance between populations connected by immigration.
2. Reference patches cannot be linked by migration to the contaminated patch. If connected, the reference patch can be affected by the toxicant.
3. Multiple discrete outcomes can occur from the same set of initial conditions.
4. Small differences in initial population sizes can dramatically alter the frequency of outcomes. It is not only the properties of the chemical and its interaction with an organism, but the status of the population that determines relative risk.

The properties of complex systems and the interconnected spatial structure of ecological systems prevent the existence of a reference or control site in ecological studies. Assumptions that ideal states exist or that pristine areas are models of uncontaminated sites are also in conflict with what are now understood as the properties of ecological systems. In other words, there is no site that shares the same history and spatial relationships as Port Valdez or any of our other study sites. However, there are two clear alternatives to the reference site model that can be employed: the use of gradients within the environment and the defining of reference conditions based upon stakeholder values.

Gradients

Contamination is just one gradient among many in a landscape. In order to understand causal relationships or to establish risk, it is necessary to sample across the confounding gradients within the area of interest. The capture of this information will include sampling up and down the gradient of contamination and potentially confounding gradients that also share at least some of the same geology and history of invasion and disturbance as the site under investigation. These criteria mean that sampling within the same landscape as the contaminated site is important.

Time is another important gradient. The effects of a contaminant can vary depending upon season, as can the effects of other gradients that can confound the detection of effects. In order to accurately represent and understand the effects of contaminants, multiple samples should be taken so that a temporal gradient can be established. Only in rare instances is information available to reconstruct events through time. In the Cherry Point, Washington study we used the sampling data over the last 30 years for the Pacific herring dynamics and used catch reports back to the late 19th century to estimate Puget Sound populations. In the case of Port Valdez, Alaska, some data were available since the earthquake and tsunami of the early 1960s on benthic community structure. However, few sites will have adequate databases.

The gradient approach requires that a great deal is known about the landscape of the region of the risk assessment. Ideally, it is nice to have data available on land-use, geology, hydrology, soil types, sediment composition, types of contaminants, history of disturbance, and other information when deciding upon a sampling plan. These data may not be available and this uncertainty should be reported.

As it turns out, many of the field studies of the past do incorporate a simplified gradient design. An upstream-to-downstream comparison is a simple gradient design and can at times detect the signal due to contamination. This approach worked well with high contaminant loadings so that the gradient of effects was very steep and the signal very large. Unfortunately, such designs lack the power to detect more subtle effects or may confuse effects due to other factors with those of the contamination.

The data required to set up a gradient analysis may make this approach impractical for performing screening for effects over large regions. For screening in order to identify areas of potential effects due to contamination a reference condition approach may be more suitable.

Reference Condition as an Alternative

Another approach is to create a reference condition that establishes either by field data or by consultation with stakeholders (groups of concerned individuals and organizations) the desired state of the ecological system. Both approaches have advantages and disadvantages.

Determining a reference condition by using field data should establish a gener-alized gradient by surveying the composition of the biological community over a variety of sites from minimal human activity to highly impacted. It must also be representative of the variety of habitat types or landforms existing within the area being managed. The representative areas sampled provide multivariate description can be constructed for each habitat or landform type. This multivariate description is the reference condition or, more to the point, a reference gradient. In this manner the variability of the ecological system that is characteristic of each type of site can be represented.

However, it is difficult to assign causality if only minimally impacted sites are sampled or if the sampling of other types of gradients is underrepresented. Sampling of sites with known impacts from a variety of known stressors would be more useful in establishing causality as such a process would approach the gradient design. Unfortunately, sampling to such an extent may not be possible given available time and resources.

Stakeholders can also define a reference state. Such a multivariate space would be bounded by the limits that stakeholders placed on the acceptable conditions for a particular site. It is also possible to construct such a site by having the stakeholders identify current sites that meet their conditions, and then sampling those sites to define an acceptable community structure. In this use of a stakeholder-defined reference state the goal is not necessarily to identify causality due to contaminants, but to identify those sites that require some form of management in order to meet the goals of the stakeholders.

The stakeholder-defined reference state approach was used in a number of sites described in this book. The Willamette–McKenzie watershed in Oregon has clearly defined management goals as set by a variety of stakeholders. All of the sites had management goals associated with them (with a variety of detail included). In few cases, the system as it existed before human impact was considered as a viable reference state.

With this background it is now appropriate to introduce the structure of this volume. It has two principal sections: description of the approach and the case studies.

THE APPROACH

The Relative Risk Model and Decision Making

Chapter 2, "Introduction to the Regional Risk Assessment Using the Relative Risk Model," is a description of the basic RRM and its application to regional risk assessment. The relationship between this approach and the standard USEPA framework is presented, as well as the ten steps in performing a complete ecological risk assessment. A comparison table summarizes the studies to date.

Chapter 3, "Interaction between Risk Assessors, Decision Makers, and Stakeholders at the Regional Scale," describes a critical aspect of any regional risk assessment, the communication between the risk assessor and the decision maker. The RRM approach is very dependent upon the determination of assessment endpoints both in its initial assessment and later in its use as a model for alternative management strategies.

CASE STUDIES

The next chapters present case studies in approximate chronological order, starting at Port Valdez, Alaska and finishing with the current work at Cherry Point, Washington. Although each chapter was written recently, the overview represents the state of the art at the time of the original study.

Chapter 4 is a summary of the first RRM regional risk assessment. The fundamental method employed and the complete set of rankings used in the process are printed for the first time. This process set the stage for the further development of the method.

In Chapter 5, our second regional-scale ecological risk assessment is detailed. This study is noteworthy because of the method of extracting stakeholder values, the conflicting assessment endpoints, and the first intensive use of geographical information systems as one of the study tools.

Codorus Creek in Pennsylvania is the subject of the next two chapters. In Chapter 6, the previously published risk assessment is summarized and the use of ecological community structure as a confirmatory tool of the risk hypothesis is detailed. The Codorus Creek studies represent our first study to test the patterns that constitute the risk assessment hypothesis with an analysis of fish and macroinvertebrate structure. Thomas (Chapter 7) then uses the RRM as an analysis tool for exploring management options within the Codorus Creek watershed.

Chapter 8, "Developing a Regional Ecological Risk Assessment: A Case Study of a Tasmanian Agricultural Catchment," was performed in the same time period as the Willamette and Codorus Creek studies. This was the first use of the RRM outside North America. Walker et al. (2001) describe an agricultural system very different in landuses compared to the previous studies.

Chapter 9, "Establishing Conservation Priorities in a Rain Forest Reserve in Brazil," deals with an enormous diversity of habitats. Not only is a rain forest part of the landscape, but so are cave environments and their ecological systems. This risk assessment is one part of a series of studies published on this Brazilian rain forest.

A smaller watershed is the subject of Chapter 10. This watershed encompasses agricultural lands, mines, shopping malls, suburbia, and flows into Bellingham Harbor. This study helped to focus decision makers upon a variety of stakeholder values as well as providing a comparison of management options within the watershed.

Chapters 11 through 13 provide descriptions of our research program at Cherry Point, Washington. These studies were initiated by the decline of the Pacific herring run that was once one of the largest in the region. Chapter 11 provides the history of the region, as well as a set of risk hypotheses for the decline of the Pacific herring. This report summarizes the prevailing ideas and approaches to understanding this region at the time our studies were initiated. Chapter 12 introduces the application of Monte Carlo analysis tools to the relative risk model. This chapter complements previously published work on the area. Chapter 13 takes another look at the Cherry Point region by using assessment endpoints in addition to the importance of Cherry Point herring for the area. The chapter explores the use of alternative habitat models in the prediction of relative risk for the Cherry Point area and its associated watersheds. Monte Carlo analysis is applied to the prediction of risk in a manner similar to that for conventional ecological risk assessments. Compared to previously published reports, we have taken the opportunity to present a more complete description of the process with an emphasis on understanding the uncertainties associated with the site.

As the editor of this volume and codeveloper of the RRM, this book represents for me the end of the beginning in the application of this approach to managing large-scale ecological systems. The future should be interesting.

REFERENCES AND SUGGESTED READINGS

Cambel, A.B. 1993. *Applied Chaos Theory: A Paradigm for Complexity*, Academic Press, Boston, MA.

Cook, R.B., Suter, G.W., II, and Sain, E.R. 1999. Ecological risk assessment in a large river–reservoir: 1. Introduction and background, *Environ. Toxicol. Chem.*, 18, 581–588.

Cormier, S.M., Lin, E.L.C., Millward, M.R., Schubauer-Berigan, M.K., Williams, D.E., Subramanian, B., Sanders, R., Counts, B., and Altfater, D. 2000. Using regional exposure criteria and upstream reference data to characterize spatial and temporal exposures to chemical contaminants, *Environ. Toxicol. Chem.*, 19, 1127–1135.

Cormier, S.M., Smith, M., Norton, S., and Neiheisel, T. 2000. Assessing ecological risk in watersheds: A case study of problem formulation in the Big Darby Creek Watershed, Ohio, USA, *Environ. Toxicol. Chem.*, 19, 1082–1096.

Hunsaker, C.T., Graham, R.L., Suter, G.W., II, O'Neill, R.V., Barnthouse, L.W., and Gardner, R.H. 1990. Assessing ecological risk on a regional scale, *Environ. Manage.*, 14, 325–332.

Landis, W.G. and McLaughlin, J.F. 2000. Design criteria and derivation of indicators for ecological position, direction and risk, *Environ. Toxicol. Chem.*, 19, 1059–1065.

Landis, W.G. and McLaughlin, J.F. 2001. If not recovery, then what?, in *Environmental Toxicology and Risk Assessment: Science, Policy and Standardization Implications of Environmental Decisions*, tenth volume, ASTM STP 1403, Greenburg, B.M., Hull, Roberts, M.H., Jr., and Gensemer, R.W., Eds., American Society of Testing and Materials, West Conshohocken, PA, pp. 283–292.

Landis, W.G. and McLaughlin, J.F. 2003. Establishing specifications of ecological indicators for the prediction of sustainability, in *Managing for Healthy Ecosystems*, Rapport, D.J., Lasley, W.L., Rolston, D.E., Nielsen, N.O., Qualset, C.O., and Damania, A.B., Eds., Lewis Publ., Boca Raton, FL, pp. 243–254.

Landis, W.G. and Wiegers, J.A. 1997. Design considerations and a suggested approach for regional and comparative ecological risk assessment, *Hum. Ecol. Risk Assess.*, 3, 287–297.

Landis, W.G. and Yu, M.-H. 2004. *Introduction to Environmental Toxicology*, 3rd ed., CRC Press, Boca Raton, FL.

Landis, W.G., Matthews, R.A., and Matthews, G.B., 1996. The layered and historical nature of ecological systems and the risk assessment of pesticides, *Environ. Toxicol. Chem.*, 15, 432–440.

Landis, W.G., Matthews, R.A., and Matthews, G.B. 1995. A contrast of human health risk and ecological risk assessment: risk assessment for an organism versus a complex non-organismal structure, *Hum. Ecol. Risk Assess.*, 1, 485–488.

Landis, W.G., Matthews, G.B., Matthews, R.A., and Sergeant, A. 1994. Application of multivariate techniques to endpoint determination, selection and evaluation in ecological risk assessment, *Environ. Toxicol. Chem.*, 12, 1917–1927.

Matthews, R.A., Landis, W.G., and Matthews, G.B. 1996. Community conditioning: an ecological approach to environmental toxicology, *Environ. Toxicol. Chem.*, 15, 597–603.

McLaughlin, J.F. and Landis, W.G. 2000. Effects of environmental contaminants in spatially structured environments, in *Environmental Contaminants in Terrestrial Vertebrates: Effects on Populations, Communities, and Ecosystems*, Albers, P.H. et al., Eds., SETAC, Pensacola, FL.

Obery, A. and Landis, W.G. 2002. Application of the relative risk model for Codorus Creek watershed relative ecological risk assessment: an approach for multiple stressors, *Hum. Ecol. Risk Assess.*, 8, 405–428.

O'Neill, R.V., Hunsaker, C.T., Jones, K.B., Riitters, K.H., Wickham, J.D., Schwartz, P.M., Goodman, I.A., Jackson, B. L., and Baillargeon, W.S. 1997. Monitoring environmental quality at the landscape scale, *BioScience,* 47, 513–519.

Spromberg, J.A., Johns, B.M., and Landis, W.G. 1998. Metapopulation dynamics: indirect effects and multiple discrete outcomes in ecological risk assessment, *Environ. Toxicol. Chem.,* 17, 1640–1649.

Suter, G.W., II. 1990. Endpoints for regional ecological risk assessments, *Environ. Manage.,* 14(1), 9–23.

Walker, R., Landis, W.G., and Brown, P. 2001. Developing a regional ecological risk assessment: a case study of a Tasmanian agricultural catchment, *Hum. Ecol. Risk Assess.,* 7, 417–439.

Wiegers, J.K., Feder, H.M., Mortensen, L.S., Shaw, D.G., Wilson, V.J., and Landis, W.G. 1998. A regional multiple stressor rank-based ecological risk assessment for the fjord of Port Valdez, AK, *Hum. Ecol. Risk Assess.,* 4, 1125–1173.

CHAPTER 2

Introduction to the Regional Risk Assessment Using the Relative Risk Model

Wayne G. Landis and Janice K. Wiegers

CONTENTS

1-56670-655-6/05/$0.00+$1.50
© 2005 by CRC Press LLC

INTRODUCTION

Since 1997 the relative risk model (RRM) proposed by Wiegers and Landis (Landis and Wiegers 1997; Wiegers et al. 1998) has been used at a variety sites to generate regional risk hypotheses on a variety of scales. These scales have ranged from an urban watershed a few square kilometers in size, to a Brazilian rain forest, and to coastal marine areas. The studies also incorporate multiple sources of multiple stressors with a variety of endpoints that exhibit a spatial and temporal distribution. The purpose of this chapter is to define regional risk assessment, present the RRM, and to briefly summarize the scope and results of the studies conducted up until the fall of 2003.

REGIONAL RISK ASSESSMENT DEFINED

Ecological risk assessment calculates the probability of an impact to a specified set of assessment endpoints over a defined period of time. In the risk assessment of chemicals, exposure and effects are estimated and the probability of the intersection of those functions calculated. Impacts typically considered are mortality, chronic physiological impacts, and reproductive effects. Most often these risk assessments deal with single chemicals in such classic cases as pesticides, herbicides, organic solvents, metals, polychlorinated biphenyls, and dioxins. Most often the risk assessments dealt with only one or a few biological endpoints.

During the 1990s there was an effort to expand ecological risk assessment to more accurately reflect the reality of the structure, function, and scale of ecological structures. Hunsaker, O'Neill, Suter and colleagues (Hunsaker et al. 1990; Suter 1990; O'Neill et al. 1997) formulated the idea of performing regional risk assessments at a landscape scale. There have been attempts to perform risk assessment based upon the classical U.S. Environmental Protection Agency (USEPA) paradigm, but each has had limitations (Cook et al. 1999; Cormier et al. 2000) imposed by a risk assessment framework originally designed for single chemicals and receptors. A principal difficulty is the incorporation of the spatial structure of the environment and the inherent presence of multiple stressors.

We (Landis and Wiegers 1997; Wiegers et al. 1998) adopted a definition that naturally incorporates multiple stressors, historical events, spatial structure, and multiple endpoints. Our working definition of a regional-scale risk assessment is:

> A risk assessment deals at a spatial scale that contains multiple habitats with multiple sources of multiple stressors affecting multiple endpoints and the characteristics of the landscape affect the risk estimate. Although there may only be one stressor of concern, at a regional scale the other stressors acting upon the assessment endpoints are to be considered.

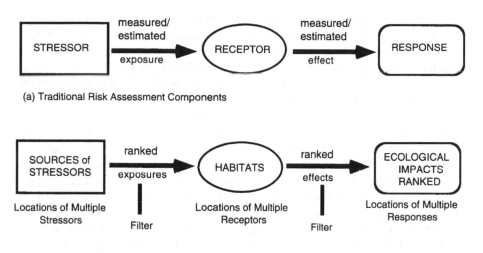

(a) Traditional Risk Assessment Components

(b) Regional Relative Risk Assessment Components

Figure 2.1   Comparison of traditional risk assessment to regional relative risk assessment.

FRAMEWORK OF THE RELATIVE RISK MODEL

The framework for the RRM for regional risk assessment was outlined by Landis and Wiegers (1997). Ecological risk assessment (EcoRA) methods traditionally evaluate the interaction of three environmental components: stressors released into the environment, receptors living in and using that environment, and the receptor response to the stressors (Figure 2.1a). Measurements or estimates of exposure and effect quantify the degree of interaction between these components. At a single contaminated site, especially where only one stressor is involved, the connection of the exposure and effect measurements to the assessment endpoints can be relatively simple. However, in a regional multiple stressor assessment, the number of possible interactions increases dramatically. Stressors arise from diverse sources, receptors are often associated with a variety of habitats, and one impact may lead to additional impacts. A complex background of sets of natural stressors and effects further clouds the picture.

Expanding an assessment to cover a region requires consideration of larger-scale regional components: sources that release multiple stressors, habitats where the multiple receptors live, and the multiple impacts to the assessment endpoints (Figure 2.1b). The three regional components are analogous to the three traditional components, but the emphasis is on location and groups of stressors, receptors, and effects.

Traditional risk assessment estimates the level of exposure and effect to calculate risk. However, exposure and effect cannot be directly measured unless a specific stressor and a specific receptor are identified. At a regional level, stressors and receptors can be represented as groups: a source as a group of stressors, a habitat as a group of receptors, and an ecological impact as a group of receptor responses. These combinations involve the use of a variety of distinctly different measurements.

For example, the measurement of a polychlorinated organic compound will results in units, mg/L, distinctly different from the occurrence of an invasive species, number of organisms/m^2. Yet both can be present within the area of study. Impacts can be similarly varied, mortality may have to be combined with a decrease with the occurrence of nonindigenous species. It is very intractable to attempt to combine measurements taken with distinctly different units.

However, it is possible to combine these measurements based on the establishment of ranks. In this manner a concentration of a chemical that may cause a high degree of mortality can be combined with an invasion of a new species that will alter a small amount of habitat. The criteria for setting ranks are discussed later, but the crucial feature is that this approach allows the evaluation of multiple stressors being derived from multiple sources impacting a variety of species in a variety of habitats in a variety of locations.

Relative regional assessment identifies the sources and habitats in different locations of the site, ranks their importance in each location, and combines this information to predict relative levels of risk. The number of possible risk combinations resulting from this approach depends on the number of categories identified for each regional component. For example, if two source types (e.g., point discharge and fish waste) and two habitat types (e.g., the benthic environment and the water column) are identified, then four possible combinations of these components can lead to an impact. If in addition we are concerned about two different impacts (e.g., a decline in the sport fish population and a decline in sediment quality), eight possible combinations exist.

Each identified combination establishes a possible pathway to a risk in the environment. If a particular combination of components interacts or affects another, then they can be thought of as overlapping. When a source generates stressors that affect habitats important to the assessment endpoints, the ecological risk is high. A minimal interaction between components results in a low risk. If one component does not interact with one of the other two components, no risk exists. For example, a discharge piped into a deep water body is not likely to impact salmon eggs, which are found in streams and intertidal areas. In such a case, the source component (an effluent discharge) does not interact with the habitat (streams and intertidal areas), and no impact would be expected (i.e., harm to the salmon eggs). This is analogous to the overlap among the stressor, receptor, and hazard in conventional risk assessment. Impact 1 may also be due to the overlap of several sources of stressors with several habitats, all altering the risk. Integrating these combinations demonstrates that impact 1 is actually the result of several combinations of sources and habitats. To fully describe the risk of a single impact occurring, each possible route to the impact needs investigation.

Integration of these routes is not always a simple matter and is again facilitated by the use of ranks. Often, measurements of various exposure and effect levels cannot be added together to determine the overall impact to the assessment endpoint. For example, a decline in wild salmon populations can result from a combination of eggs in the spawning grounds being exposed to chemicals and increased predation when the juveniles migrate out of the port. However, chemical exposure to the eggs may also influence growth of the juvenile fish. Smaller fish are less able to avoid predation, and mortality from predation may increase beyond what would be expected if the effect to the eggs was not considered.

The RRM regional approach is a system of numerical ranks and weighting factors to address the difficulties encountered when attempting to combine different kinds of risks. Ranks and weighting factors are unitless measures that operate under different limitations than measurements with units (e.g., mg/L, individuals/cm^2) (Figure 2.2). In a complex system with a wide range of dissimilar stressors and effects, few measurements exist that are additive. For example, there is little meaning in adding toxicant concentrations to counts of the number of introduced predators in order to determine the total risk in a system. However, knowing that a particular region has both the highest concentrations of a contaminant and the most introduced predators is useful in a decision-making process.

The next sections take this basic approach and describe the steps in conducting a regional relative risk assessment, from problem formulation to risk communication.

THE 10 STEPS OF THE RELATIVE RISK MODEL FOR REGIONAL RISK ASSESSMENT

The previous reviews of the application of the RRM have led to the formulation of ten procedural steps that formalize the process. The process can also generate three specific outputs useful in the decision-making process.

The procedural steps are

1. List the important management goals for the region. What do you care about and where?
2. Make a map. Include potential sources and habitats relevant to the management goals.
3. Break the map into regions based upon a combination of management goals, sources, and habitats.
4. Make a conceptual model that links sources of stressors to the receptors and to the assessment endpoints.
5. Decide on a ranking scheme to allow the calculation of relative risk to the assessment endpoints.
6. Calculate the relative risks.
7. Evaluate uncertainty and sensitivity analysis of the relative rankings.
8. Generate testable hypotheses for future field and laboratory investigation to reduce uncertainties and to confirm the risk rankings.
9. Test the hypotheses listed in Step 8.
10. Communicate the results in a fashion that portrays the relative risks and uncertainty in a response to the management goals.

These ten steps correspond to the portions of the ecological risk assessment framework as depicted in Figure 2.3. The first four steps of the RRM correspond to the initial segments of the framework, especially problem formulation. These initial steps largely determine the success of the risk assessment. Steps 4, 5, and 6 are closely related and do not fit cleanly into conventional framework. The conceptual

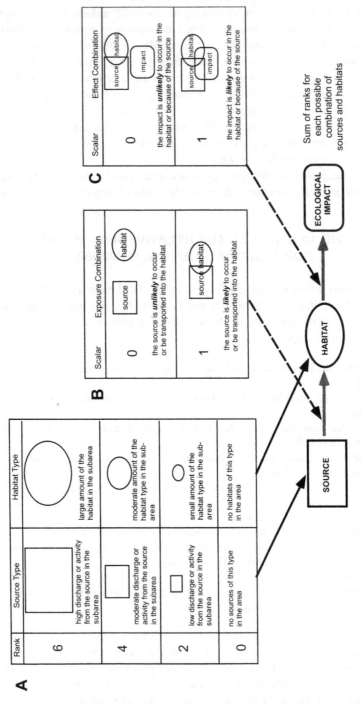

Figure 2.2 The application of ranks and filters in the RRM scheme.

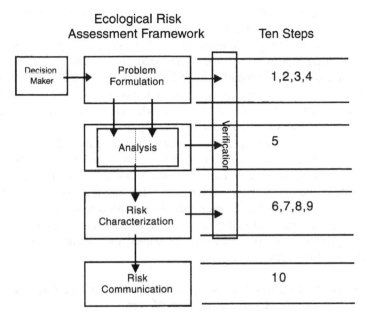

Figure 2.3 Relationship of the ten steps in the RRM to the classic ecological risk assessment paradigm.

model is based upon knowledge of source–stressor habitat–effects linkages. Determination of the ranking scheme incorporates a large quantity of data generated on the amounts of stressors, habitats, and what knowledge is available on potential outcomes. Once the conceptual model and ranking scheme are established the actual calculation is straightforward. Analysis of uncertainty and sensitivity and generation of testable hypotheses are the more difficult steps that most closely correspond to risk characterization. Testing the hypotheses corresponds to the verification step and should be incorporated whenever possible.

Step 10 corresponds to risk communication and is comprised of three outputs.

1. Maps of the risk regions with the associated sources, landuses, habitats, and the spatial distribution of the assessment endpoints.
2. A regional comparison of the relative risks, their causes, the patterns of impacts to the assessment endpoints, and the associated uncertainty. These regional comparisons and estimates of the contribution of each source and stressor create a spatially explicit risk hypothesis.
3. A model of source–habitat–impact that can be used to ask what-if questions about different scenarios that are potential options in environmental management.

These outputs summarize the data and provide risk assessments and a tool for the examination of different risk scenarios. These outputs facilitate communication and decision making for the environmental managers. The next section describes each of the ten steps and the three outputs.

The first four steps are critical to performing a regional ecological risk assessment and are the foundation of a useful risk assessment that can be applied to the decision-making process and to long-term environmental management. These steps should involve close interaction with all of the interested parties. The parties include the regulators, the regulated community, the stakeholders comprised of private citizens and nongovernmental organizations, and the risk assessors. There are likely to be environmental managers in the first three groups who will be involved in the decision-making process. The risk assessors need to clearly understand the decision-making needs of each of the other groups, communicate the strengths and limitations of the risk assessment process, and attempt to translate management goals stated in non-scientific terminology to features that can be quantified and evaluated. In this inter-action the role of the risk assessor is clearly not decision making, but scientific and technical support. At times the decision makers may need to know that a particular goal is not part of ecological reality, or that the field of science is not sufficiently advanced to provide predictive measures. However, the interaction is critical if a successful risk assessment is to occur.

Step 1. List the Important Management Goals for the Region. What Do You Care about and Where?

The management goals are the key to the rest of the risk assessment. Regional risk assessments are most effective when they target the decision-making needs and goals of environmental managers. It is important to identify difficult or even con-flicting goals. Decisions must be identified early in the process. Without identifying, discussing, and resolving these issues, the assessment results will not appear to be useful to managers, and in fact may not be usable for the decisions at hand.

There are four sets of interactions among the regulated community, the regula-tors, and the interested stakeholders in the decision-making process. Interaction among these three groups is expected in three forms. First, each will interact with the other two parties in a bipartite fashion. Second, all three parties must interact at the same time to clearly define the management and decision options in order to answer basic questions about the future management of the area. Third, there are also interactions between the three groups and the risk assessment team.

The role of the risk assessment team is critical. In some instances the desired uncertainty reduction is not possible due to resource limitations (Suter 1993), and some management goals are unattainable as well. While a goal may be to restore the balance of nature or to return the system to a pristine state, given our current understanding of ecological systems, neither of these goals is attainable (Landis and McLaughlin 2000). However, stakeholders envision the restoration of certain eco-logical resources to within usable limits, and these goals can be quantified and engineered.

The management goals for the fjord of Port Valdez and the Codorus Creek watershed in Pennsylvania were derived from public meetings with representatives of the various stakeholder groups. These groups included the regulated community, the regulators, interested stakeholders, and the risk assessors.

Table 2.1   Examples of Stakeholder Values for Two Sites of Regional-Scale Risk Assessments

Willamette–McKenzie River, OR	Codorus Creek, PA
River water is usable as source of drinking water	Protective water quality for aquatic ecological receptors and humans during contact or consumption
Fish from river are palatable and safe to eat	
There are sufficient numbers of desirable fish to support an active recreational and commercial fishery	Adequate water supply for drinking and waste discharge
Summer steelhead populations	Self-sustaining native and nonnative fish populations in the watershed
Spring chinook salmon populations	Adequate food availability for aquatic species
River sustains thriving populations of native fish	Available recreational land and water resources
Floodplain protection and enhancement for natural functions and values	Adequate stormwater control and treatment
Floodplain management for human health and safety	
Water quantities sustain human communities	
Maintain reservoirs for fishing, boating, and windsurfing	

In some instances, such as the Willamette–McKenzie risk assessment, a similar process may already have been performed by the appropriate stakeholder groups. In the Willamette–McKenzie study the values were derived from the Willamette Valley Livability Forum, a group established by the governor of Oregon with a charge of establishing management goals for the ecological services provided by the Willamette River and its tributaries. The process was driven by consensus for the period up to 2050. The management goals for fisheries are shown in Table 2.1. The first column lists the goals as defined by this group. The second column is the quantitative measure that we used to define this goal. In some areas there are conflicts where two desired goals appear incompatible, but the goal of the risk assessment team is to be as inclusive as possible.

As this process is completed the management goals are then placed into a spatial context with the appropriate sources and habitats.

Step 2.   Make a Map. Include Potential Sources and Habitats Relevant to the Management Goals

As an example we will use the map-making process for the Cherry Point study, but all of the studies to date incorporate a similar process. First, the potential sources within the study area are located, characterized, and placed on a map that includes the critical topological features of the system. The boundaries are set by the management goals of the decision makers, but also take into account the life history of the various endpoints. Habitat information is also plotted for the endpoints under consideration. Maps can be produced in a variety of ways; the Port Valdez study utilized conventional maps scanned into a computer and the additional information was added in a graphics program. Subsequent studies have made extensive use of geographical information systems (GIS) that have distinct advantages and disadvantages. The advantages are clearly the ability to display and analyze geographical information in a variety of formats. Unfortunately, not all spatial data are in digital form, digital data can often be expensive when it does exist, and digital data are kept in a variety of projections which take time to combine. Uncertainty related to geographical information is also an issue that will be discussed in Step 7.

The next step is to combine management objectives, source information, and habitat data into geographically explicit portions that can be analyzed in a relative manner.

Step 3. Break the Map into Regions Based upon a Combination of Management Goals, Sources, and Habitats

The next step is the creation of risk regions that delineate the boundaries of the areas for which risks will be calculated. This map is the basis of the rest of the analysis because risks are all relative based upon the delineated regions. The map is also based upon possible pathways of exposure in a spatial sense to the locations where habitat can be found for the assessment endpoints. In this regard it may be very important to follow fate of the water, groundwater, soil, and air within the landscape to ensure that appropriate sources, stressors, and habitats are incorporated into a risk region. The chapters that follow in this text provide a variety of methods of deriving risk regions.

Step 4. Make a Conceptual Model that Ties the Stressors to the Receptors and to the Assessment Endpoints

The conceptual model delineates the potential connections between sources, stressors, habitat, and endpoints that will be used in each risk region. An example of such a conceptual model for hypothetical regional-scale mining and smelting site is presented in Figure 2.4 and was constructed by E. Hart Hayes. The site is in a heavily forested area along a major river, with dams, transportation corridors, and other activities occurring in the same region. The conceptual model is an extension of the basic framework for a regional risk assessment with sources providing stressors into particular habitats. In this instance the habitats are broadly defined as terrestrial and aquatic to capture the exposure pathways and location within the region of our endpoints. There are numerous interconnected endpoints both to show the valued ecosystem components and to illustrate the interdependence and potential indirect effects.

In cases (such as this illustration) where metals can be assumed to be the principal contaminant, it is important to incorporate all of the confounding stressors. The shaded boxes (Figure 2.4) highlight the conceptual model if only metals were being considered. However, all of the endpoints are also being impacted by other stressors as well. A metals-only assessment would take the endpoints and the metals out of context.

A well-constructed and informative conceptual model places the site, the stressors, the habitats, and the effects into a regional context. Such a construction can eliminate some stressors due to the lack of exposure pathways and lead to the inclusion of confounding factors outside the original scope of the assessment.

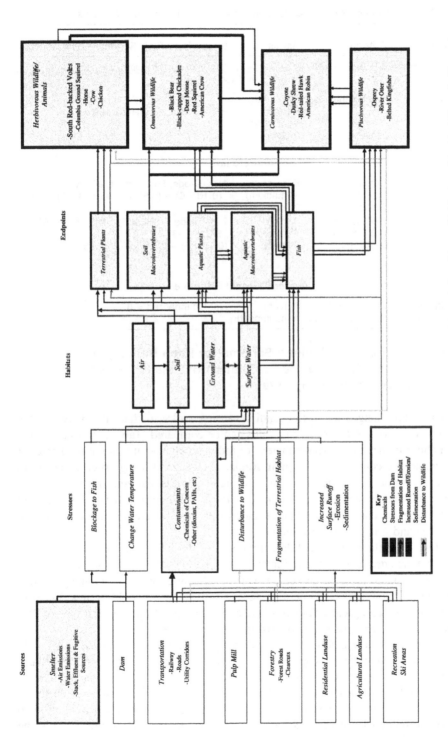

Figure 2.4 Example of a conceptual model incorporating the basic framework of the relative risk model (designed by Hart Hayes). (See color insert following page 178.)

Table 2.2   Example of Ranking Criteria for Stressors for Codorus Creek, PA

Coverage	Criteria	Ranks	Example — Risk Region 1 Rank Scores
		Landuse	
Industrial	% Industrial		
	< 1	6 (high)	< 1% Industrial = Rank of **2**
	< 1–2	4 (medium)	
	2–16	2 (low)	
		Soil Erosion	
Vegetation	Crops	6 (medium)	16% Crops, 59% Forest, 24% Grass
	Forest	4 (medium)	$(0.16 \times 6) + (0.59 \times 4) + (0.24 \times 2) = $ **3.5**
	Grass	2 (low)	
Soils	> 8% Slope	6 (high)	70% Slope > 8%, 25% Slope 3–8%, 5%
	3–8% Slope	4 (medium)	Slope 0–3
	0–3% Slope	2 (low)	$(0.70 \times 6) + (0.25 \times 4) + (0.05 \times 2) = $ **5.3**
Average			4.4
		Altered Channel Structure	
Channelization	Channelized	6 (high)	Not Channelized = **0**
	Not Channelized	0 (no impact)	

Step 5.   Decide on a Ranking Scheme for Each Source, Stressor, and Habitat to Allow the Calculation of Relative Risk to the Assessment Endpoints

This step changes data into nondimensional ranks so that effects due to the various stressors to the various endpoints can be compared (Table 2.2). Each source and habitat is ranked between subareas to indicate whether it is high, moderate, or low within the context of the region. Ranks are assigned using criteria specific to the study region. The criteria are based typically on the size and frequency of the source and the amount of available habitat. Ranks are assigned for each source and habitat type, generally on a two-point scale from 0 to 6 where 0 indicates no habitat or source and 6 is the greatest amount.

There are different means of determining the criteria for ranks. In some instances there may be adequate concentration response and fate of the stressor data available to assign ranks to a particular source. For an effluent containing one nonpersistent compound, below an EC10 could be zero, EC10 to EC30 could be low, EC30 to EC50 medium, and greater than an EC50 could be high. Typically, that type of data is not available for most stressors arising from a source.

In the chapters that follow there are many examples of ranking schemes with the criteria listed in the accompanying tables. In the case of the Port Valdez scenario (Chapter 4) these tables have been expanded beyond those of the original publication to show all the variables included in the risk assessment. In some instances clustering algorithms (Codorus Creek, Squalicum Creek, Cherry Point, etc.) were used to determine natural breaks for the ranking criteria. The details are presented in the following chapters.

Step 6. Calculate the Relative Risks

Filters determine the relationships among the risk components (source, habitat, and impact to assessment endpoints). A filter consists of the weighting factors, 0 or 1, that indicate either a low or a high probability. We have incorporated two types of filters: an *exposure filter* and an *effect filter.* The exposure filter screens the source and habitat types for the combinations most likely to result in exposures (i.e., receptors in the habitat will come into contact with stressors generated by the source). The effect filter screens the source and habitat combinations for those most likely to affect a specific assessment endpoint. The examples below describe the design of both an exposure and an effects filter.

The first step in designing an exposure filter is to determine which stressors are produced by the sources. Professional knowledge is then used to answer two sequential questions about each stressor in relation to specific source–habitat combinations:

- Will the source release or cause the stressor?
- Will the stressor then occur and persist in the habitat?

If the answer to both questions is yes, then 1 is assigned to the source–habitat combination. If the answer to either question is no, then 0 is assigned.

The design of an effect filter is similar, but a separate filter is made for each assessment endpoint. The first step in this process is to determine what type of effects is important to the specific endpoint. For instance, if maintaining crab populations is an assessment endpoint, some of the important effects to consider are toxicity, predation, and food availability. The questions asked to develop the effect filters are:

- Will the source release stressors known to cause this particular effect to the endpoint?
- Are receptors associated with the endpoint sensitive to the stressor in this habitat?

If the answer to both questions is yes, then 1 is assigned to the source–habitat combination. If the answer to either question is no, then 0 is assigned.

Integrating Ranks and Filters

Ranks and weighting factors are combined through multiplication. The results are a relative estimate of risk in each subarea. Final risk scores (RS) are calculated for each subarea by multiplying ranks by the appropriate weighting factor (W_{ij}) as indicated below.

$$RS = S_{ij} \times H_{ik} \times W_{jk} \tag{2.1}$$

where:
i = the subarea series (Region 1, 2, 3, etc.),
j = the source series (discharge ..., shoreline activity),
k = the habitat series (mudflat ..., stream mouth),

S_{ij} = rank chosen for the sources between subareas,
H_{ik} = rank chosen for the habitats between subareas,
W_{jk} = weighting factor established by the exposure or effect filter.

The results form a matrix of risk scores related to the relative exposure or effects associated with a source and habitat in each subarea. The potential risk resulting from a specific source (Equation 2.2) and occurring within a specific habitat (Equation 2.3) can be summarized for each subarea by adding the related scores,

$$RS_{source} = \Sigma(S_{ij} \times H_{ik} \times W_{jk}) \text{ for } j = 1 \text{ to } n, \tag{2.2}$$

$$RS_{habitat} = \Sigma(S_{ij} \times H_{ik} \times W_{jk}) \text{ for } k = 1 \text{ to } n. \tag{2.3}$$

Step 7. Evaluate Uncertainty and Sensitivity Analysis of the Relative Rankings

Uncertainty needs to be accounted for and tracked in the risk assessment process. Narratives can list the factors that introduced uncertainty into the assessment process. It is also possible to examine uncertainty in a variety of quantitative means including the Monte Carlo process employed to provide a range of values.

In the case of the Codorus Creek risk assessment (Obery and Landis 2002), three sensitivity evaluations were performed to examine uncertainty of the model. These methods were single-component analysis, exposure pathway analysis, and random component analysis. Single-component analysis consisted of standardizing individual stressors in each of the risk regions to test the sensitivity of the model. Exposure pathway analysis consisted of altering pathways with weak relationships in the conceptual model warranting inclusion or exclusion in the evaluation. This uncertainty analysis was warranted because only pathways demonstrating a strong relationship between the stressors–habitats and habitat–endpoints were evaluated during the risk characterization. Random component analysis evaluated model bias by assigning random numbers during 20 simulations to stressors and habitats for each risk region. Microsoft Excel® was used to generate a table of random numbers from an even distribution of values from 0 to 6.

To quantify a range of realistic conditions in the watershed, maximum and minimum reasonable ranks were determined. Landuse, surface erosion, wastewater discharge, macroinvertebrate habitat, riparian habitat, and urban park habitat ranks are believed to represent site-specific conditions; however, ranking methods for streambank development, surface runoff, altered flow rates, and fish habitat may not be as representative of actual conditions. These stressors and habitats were altered using best professional judgment to reflect reasonable maximum and minimum scenarios.

Box plots are generated from the results of these uncertainty analyses to illustrate the risk range, and the Codorus Creek analysis is presented in Figure 2.5. The single-component analysis demonstrated that changing a single rank to the same value produces total risk ranks of relatively equal magnitude, demonstrating that no single area is sensitive. This analysis can also be extrapolated to show the impact of using

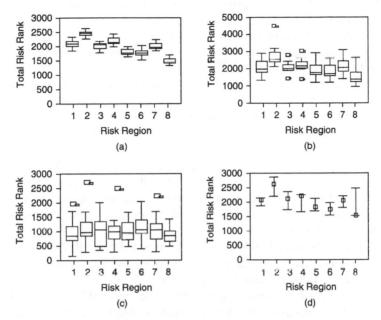

Figure 2.5 Uncertainty analysis box plots for the Codorus Creek risk assessment: (a) uncertainty analysis altering single components, (b) uncertainty analysis altering pathways of exposure, (c) uncertainty analysis using random numbers, and (d) results with reasonable maximum and minimum ranks. (After Obery, A. and Landis, W.G., *Hum. Ecol. Risk Assess.*, 8, 1779–1803, 2002. With permission.)

arithmetic mean ranks. The impact of assessing all macroinvertebrate habitats as high quality was also evaluated. When macroinvertebrate habitat was excluded from the assessment, ranks in three regions increased by a single level. Specifically, Regions 1 and 4 changed from a medium to a high rank, and Region 6 changed from a low to a medium rank. The results of excluding macroinvertebrate habitat from the assessment illustrated that the original ranks might be underestimating the risks in Regions 1, 4, and 6. This finding is consistent with the assessment results as Regions 1 and 6 have the highest habitat ranks, and Region 4 has moderately high amounts of habitats and stressors, which together cause elevated risks.

Exposure pathway analysis demonstrated a wide variance in the results. For example, impacts to riparian habitat may be doubled from the inclusion of landuse with the remaining stressors. The uncertainty assessment showed that exclusion of landuse when evaluating riparian habitat resulted in a 3 to 14% decrease of total risk ranks; however, it did not result in any ranks clustered into different risk categories. Similarly, a 5 to 21% increase of total risk resulted from the inclusion of water quality, fish populations, and food for fish populations assessment endpoints impacted in urban park habitat. This evaluation determined that Region 6 would change from a low to a medium rank. When all endpoints were considered to be complete pathways, ranks remained the same.

Random component analysis evaluated model bias by assigning random numbers to stressors and habitats for each risk region. From the 20 simulations, it was concluded that random values produced random results (Figure 2.5). No patterns were demonstrated from the exercise.

Maximum and minimum risk ranges are also illustrated in Figure 2.5. This assessment indicated that Regions 1, 4, and 5 ranks may overestimate the risk and Region 8 ranks may underestimate the risk.

Because Region 2 risk was substantially higher than the other regions, ranks were also classified using four natural breaks (i.e., very high, high, medium, and low) instead of three natural breaks. Results indicated Region 2 as very high risk, Regions 1, 3, 4, and 7 as high risk, Regions 5 and 6 as medium risk, and Region 8 as low risk. Using this ranking scheme, reasonable maximum ranks change for Regions 1, 4, and 7 from high to medium, Region 8 from low to high, and Region 6 from medium to low. This result is consistent with the findings that Region 8 risk may be underestimated. Minimum ranks using four natural breaks show Region 2 as very high risk, Regions 1 and 7 as high risk, Regions 3, 4, and 5 as medium risk, and Regions 6 and 8 as low risk.

More recently Monte Carlo analysis has been added to the RRM process. In risk assessment, Monte Carlo uncertainty analysis combines assigned probability distributions of input variables to estimate a probability distribution for output variables. In the case of the Cherry Point regional risk assessment (Chapter 13), the input variables are the ranks and filters with medium or high uncertainty and the output variables are the risk estimates.

For the Monte Carlo uncertainty analysis, we first assign designations of low, medium, or high uncertainty to each source and habitat rank, exposure, and effects filter based on data quality and availability. We assign discrete probability distributions to ranks and filters with medium and high uncertainty. The details of the process of assigning distributions to the variables are covered in the Cherry Point risk assessment chapter.

Using Crystal Ball® 2000 software as a macro in Microsoft® Excel 2002, the Monte Carlo simulations are run for 1000 iterations and output distributions for each subregion, source, habitat, and endpoint risk prediction are calculated. These distributions show a range of probable risk estimates associated with each point estimate.

Figure 2.6 illustrates the distributions for the relative risk to the assessment endpoints associated with Cherry Point (Chapter 13). A terrestrial species, the great blue heron, and juvenile Dungeness crab are the endpoints clearly at the highest risk even given the associated uncertainties. The remainder of the endpoints are at lower risks and with the given uncertainties are essentially at the same risk level. Although the distributions are depicted as continuous in the illustrations for clarity, it should be noted that the RRM is a discrete multinomial model.

Step 8. Generate Testable Hypotheses for Future Field and Laboratory Investigation to Reduce Uncertainties and to Confirm the Risk Rankings

The combination of Steps 6 and 7 produces risk hypotheses that constitute patterns in the landscape and risks to the endpoints. These hypotheses can be tested if there are adequate field-related data. A risk assessment should be able to provide

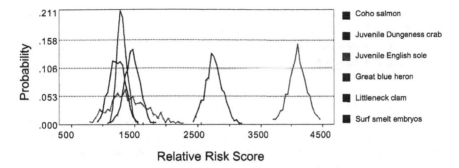

Figure 2.6   Monte Carlo output distributions for risk to assessment endpoints. (After Hart Hayes, E. and Landis, W.G., *Hum. Ecol. Risk Assess.*, 10, 299–325, 2004.)

predictions that can be tested using a variety of methods. It may not be possible to perform landscape-scale experimental manipulations, but it is clearly possible to make predictions about patterns that should already exist. The hypothesis to be tested may be a subhypothesis of the overall risk estimation that is clearly testable. Being able to test and confirm at least part of the hypotheses generated by the risk assessment should increase the confidence of the risk assessors, stakeholders, and decision makers in using the results for environmental management.

Step 9.   Test the Hypotheses Listed in Step 8

Hypotheses can be tested using a variety of field, mesocosm, or laboratory test methods. In an ideal situation it should be possible to make predictions based upon known concentrations and then sample that field site in order to confirm effect or no-effect. It may be necessary to rework the risk assessment in order to reduce uncertainty, or a stressor–habitat–effect linkage may be incorrect. Testing the risk predictions allows feedback into the assessment process, improving future predictions.

In the Valdez assessment (Chapter 4) a variety of chemical data that were not included in the original assessment were used to test the assumptions about exposure. These hypotheses were largely confirmed.

Obery, Thomas, and Landis (Chapter 6) used the information on fish and macrobenthic community structure to compare the patterns within the Codorus Creek watershed to the patterns of risk derived from the risk assessment. In this instance the patterns of risk and patterns within the biological communities matched.

Clearly, it is possible to test risk hypotheses with many implications for future monitoring programs and the adaptive management of risk.

Step 10.   Communicate the Results in a Fashion that Portrays the Relative Risks and Uncertainty in a Response to the Management Goals

The risk assessment process, no matter how scientifically valid, is still not useful unless the results are clearly communicated to the stakeholders and decision makers

who commissioned the study. Duncan (Chapter 3) discusses and stresses the importance of this activity. A variety of tools can be used.

Three outputs have been found that are particularly useful in communicating the
results of the risk assessment.

1. Maps of the risk regions with the associated sources, landuses, habitats, and the
   spatial distribution of the assessment endpoints.
2. A regional comparison of the relative risks, their causes, the patterns of impacts
   to the assessment endpoints, and the associated uncertainty. These regional comparisons and estimates of the contribution of each source and stressor create a
   spatially explicit risk hypothesis.
3. A model of source–habitat–impact that can be used to ask what-if questions about
   different scenarios that are potential options in environmental management. This
   type of process has now been performed for Codorus Creek (Chapter 6) and
   Squalicum Creek (Chapter 10), and I refer the reader to these chapters for details.

OVERVIEW OF THE RELATIVE RISK MODEL STUDIES

There are nine study sites that have been examined using the RRM, and these
are summarized in Table 2.3. The studies are presented in approximate chronological
order. There always has existed a great deal of overlap in the timelines of each study.
Except for the Leaf River in Mississippi and the Loa Watershed in Chile, each of
these sites is represented by a chapter in this volume.

The size of the study sites ranges from 62 km^2 for the Squalicum Creek Watershed
to 33,570 km^2 for the Loa Watershed. Port Valdez and Cherry Point are marine sites
and the remainder of the sites are comprised of freshwater watersheds or saltpans.
The sites are in the Americas except for Mountain River, Tasmania.

Endpoints are not as varied as the diversity of the sites. Water quality, recreational
uses, subsistence, sport fishing, and persistence of the aquatic environment are
endpoints important in each site. The persistence of macroinvertebrate communities
often is shown as an important endpoint, especially as they contribute to the persistence of the fish populations. In sites in the western United States, native salmonids
are always seen as an important part of the ecological system for preservation, but
each location seems to have its own representative fish endpoint.

There has been a great deal of methods development during each of these studies.
The basic approach was set by Wiegers et al. (1998, Chapter 4), and that foundation
has remained intact. GIS have become a critical part of this approach. GIS is so
important now that if digital data are not available, maps are scanned in and converted
to electronic form. The assessments were prospective until we were asked to examine
the causes of the decline of the Cherry Point Pacific herring.

The analysis of the causal factors leading to a decline of the Pacific herring stock
at Cherry Point, Washington (Chapter 11) was the first time that the RRM was used
as a tool to examine causation. Essentially, the RRM acts as a framework for a
weight-of-evidence process. This analysis demonstrated that the causes of the decline
were not specific to the Cherry Point region, and that Pacific herring are a poor
indicator of the status of that region of coastal Washington (Landis et al. 2004). Hart

Table 2.3 Summary of the Risk Assessments Using the Relative Risk Model

Site Location	Size	Risk Assessment Endpoints	Methods	Uncertainty	Highlights	Lessons/ Improvements	References
Port Valdez, Alaska	94.5 mi² (151.2 km²)	Water quality, sediment quality, decrease in hatchery salmon returns, population declines of bottom fisheries, declines in wild populations of anadromous fish, decreased bird populations, decreased food for wildlife populations	RRM for risk assessment. Confirmation by comparing chemistry data to benchmarks and by using a predictive model to estimate toxicity due to 10 hydrocarbon compounds found in the sediments; mapping using conventional techniques	Detailed written description used to document uncertainty; sensitivity analysis performed on the RRM; random iterations performed on the ranks of sources and stressors to observe range of outcomes	Specific risks applied on a region by region basis; area with highest risk (mudflats) had been overlooked in previous assessments	Development of the RRM, including methods of evaluating uncertainty and confirmation of the risk predictions	Landis and Wiegers 1997; Wiegers et al. 1998
Willamette–McKenzie watersheds, Oregon	1351 mi² (2179 km²)	Salmonids: spring chinook, rainbow and cutthroat trout, summer steelhead; other assessment endpoints identified and used to demonstrate conflicts with salmonid endpoints	RRM for risk assessment; Arc View® and Arc Info® used to compile and compare environmental data and to produce maps; risk confirmation by comparing patterns of water quality and toxicity to that of the risk assessment	Same as Port Valdez	RRM predicted two general areas of relatively high risk: the uppermost segment and the mouth of the McKenzie River; although the scores were similar, the underlying causes were very different	Implemented the use of GIS into the development of the RRM; used stakeholder documents as a means of setting assessment endpoints	Landis et al. 1998, 2000; Luxon, 2000; Luxon and Landis, Chapter 5, this volume

Table 2.3   Summary of the Risk Assessments Using the Relative Risk Model (continued)

Site Location	Size	Risk Assessment Endpoints	Methods	Uncertainty	Highlights	Lessons/ Improvements	References
Mountain River, Tasmania, Australia	190 km²	Water quality, maintenance of adequate stream flow, maintenance or increase of native streambank vegetation and reduction of weed density to less than 10% of groundcover; maintenance of primary industries, landscape aesthetics, and a good residential environment	RRM for risk assessment; Arc View and Arc Info used to compile and compare environmental data and to produce maps	Same as Port Valdez	Initial study on a broad agricultural area, first transfer of the RRM to an outside group	Improved use of GIS, created own computer database from scanned materials	Walker et al. 2001; Chapter 8, this volume
PETAR, Brazil	1000 km²	Self-sustaining epigean (surface) and hypogean (cave) aquatic fauna	RRM for risk assessment; Arc View and Arc Info used to compile and compare environmental data and to produce maps; introduction of the weighting system for stressor to account for differences in the amounts of stressors emitted from the various sources	Same as Port Valdez; incorporates upstream contribution to risk downstream	Assessment of both above and belowground habitats by mapping geological regions favorable to cave formation; applicability of results in the management of the natural reserve and in the guidance of site-specific investigations	Inclusion of data collected on the site, first use in a rain forest site	Moraes et al. 2002; Chapter 9, this volume

Location	Area	Endpoints	Methods		Results	Comments	Reference
Codorus Creek, PA	719 km² (278 mi²)	Water quality, water supply, self-sustaining native and nonnative fish populations, adequate food supply for aquatic species, recreational land and water resources, stormwater control and treatment	RRM for risk assessment; Arc View and Arc Info used to compile and compare environmental data and to produce maps; confirmation of fish and macroinvertebrate community structure by multivariate analyses	Same as Port Valdez	Urbanization was the greatest risk factor within the watershed; patterns of risk were confirmed by the field research and multivariate analysis	Use of multivariate methods to evaluate risk predictions; application of predictive modeling	Obery and Landis 2002
Squalicum Creek, WA	62 km²	*Abiotic endpoints*: flood control, adequate land and ecological attributes for recreational uses; *biotic endpoints*: viable nonmigratory coldwater fish populations, life cycle opportunities for salmonids, viable native terrestrial wildlife species populations, adequate wetland habitat to support wetland species populations	RRM for risk assessment: Arc View and Arc Info used to compile and compare environmental data and to produce maps	Same as Port Valdez	RRM was adapted for a small watershed in a rapidly urbanizing environment	Application of the RRM in a very small and urbanized watershed; direct cooperation with the planners and managers of Squalicum Creek	Chen 2002; Chen and Landis, Chapter 10, this volume

Table 2.3 Summary of the Risk Assessments Using the Relative Risk Model (continued)

Site Location	Size	Risk Assessment Endpoints	Methods	Uncertainty	Highlights	Lessons/ Improvements	References
Cherry Point, WA	715 km²	Cherry Point Pacific herring run for retrospective assessment; prospective assessment includes coho salmon, juvenile English sole and surf smelt embryos, juvenile Dungeness crab, adult littleneck clam, and great blue heron	Retrospective and prospective risk assessment using the RRM approach; Arc View and Arc Info used to compile and compare environmental data and to produce maps	Initial approach similar to Port Valdez; Monte Carlo techniques used to evaluate uncertainty and sensitivity; examined the impact of different assumptions concerning the extent of habitat type and effect upon the assessment population in determining risk	Retrospective study pointed to the influence of factors beyond the Cherry Point region as the cause of the herring decline; other endpoints adopted as more relevant to the management of the area; a bird, the great blue heron, was shown to be most at risk for the area; eventual development of a weight-of-evidence approach to the retrospective risk assessment with application of Monte Carlo techniques	First retrospective use of the RRM; marked the first use of Monte Carlo techniques in evaluating uncertainty in the RRM	Hart Hayes et al. 2004; Landis et al. 2004.

Leaf River, MS	5766 km² (3575 mi²)	Fish, macroinvertebrates, water quality, water quantity, recreational uses, wastewater treatment, channel modifications	Prospective risk assessment using the RRM approach at a very large scale in a watershed very different than the other studies; used field data to test the risk hypotheses; also incorporated predictive modeling in an examination of risk management schemes	Added pathway analysis to examine the sensitivity of assumptions about linkages in exposure and effects pathways to the final risk estimates	Incorporates all ten steps in a clear fashion; hypotheses tested using an analysis of community structure	Included analysis of the sensitivity of the models to the pathways, broad-scale risk assessment	Thomas 2003
Loa Watershed, Chile	33,570 km²	Aquatic life in rivers and saltpans (shallow lagoons of water rich in salts)	Retrospective risk assessment using the RRM approach; Arc View and Arc Info used to compile and compare environmental data and to produce maps	Same as Port Valdez	Largest scale assessment to date; assessment in a mining area in northern Chile using the RRM approach at a very large scale in desert conditions	Applicability of the model in a large area, but high uncertainty due to large distances between sources and habitats and possible uncompleted pathways of exposure	Hamamé 2002

Hayes (Hart Hayes and Landis 2004) performed a risk assessment using alternative endpoints that more accurately represent the status of the particular coastal area.

The two largest scale assessments are those by Thomas (2003) and Hamame (2002). They cover very different aquatic systems: the Leaf River in Mississippi and the Loa Watershed in the arid lands of Chile. These studies demonstrate that basic methodology can be used for a wide variety of scales and in very different environments.

Of course, the most sophisticated methodology is not useful if it does not address the needs of the decision makers. The next chapter discusses the critical issue of the interaction of regional-scale risk assessment with the decision-making process.

REFERENCES

Cook, R.B., Suter, G.W., II, and Sain, E.R. 1999. Ecological risk assessment in a large river–reservoir: 1. Introduction and background, *Environ. Toxicol. Chem.,* 18, 581–588.

Cormier, S.M., Lin, E.L.C., Millward, M.R., Schubauer-Berigan, M.K., Williams, D.E., Subramanian, B., Sanders, R., Counts, B., and Altfater, D. 2000. Using regional exposure criteria and upstream reference data to characterize spatial and temporal exposures to chemical contaminants, *Environ. Toxicol. Chem.,* 19, 1127–1135.

Cormier, S.M., Smith, M., Norton, S., and Neiheisel. T. 2000. Assessing ecological risk in watersheds: a case study of problem formulation in the Big Darby Creek Watershed, Ohio, USA, *Environ. Toxicol. Chem.,* 19, 1082–1096.

Hamamé, M. 2002. Regional Risk Assessment in Northern Chile Report 2002: 1. Environmental Systems Analysis, Chalmers University of Technology, Göteborg, Sweden.

Hart Hayes, E. and Landis, W. G. 2004. Regional ecological risk assessment of a nearshore marine environment: Cherry Point, WA, *Hum. Ecol. Risk Assess.,* 10, 299–325.

Hunsaker, C.T., Graham, R.L., Suter, G.W., II, O'Neill, R.V., Barnthouse, L.W., and Gardner, R.H. 1990. Assessing ecological risk on a regional scale, *Environ. Manage.,* 14, 325–332.

Landis, W.G. 2002. Uncertainty in the extrapolation from individual effects to impacts upon landscapes, *Hum. Ecol. Risk Assess.,* 8, 193–204.

Landis, W.G. and McLaughlin, J.F. 2000. Design criteria and derivation of indicators for ecological position, direction and risk, *Environ. Toxicol. Chem.,* 19, 1059–1065.

Landis, W.G. and Wiegers, J.K. 1997. Design considerations and a suggested approach for regional and comparative ecological risk assessment, *Hum. Ecol. Risk Assess.,* 3, 287–297.

Landis, W.G., Duncan, P.B., Hart Hayes, E., Markiewicz, A.J., and Thomas, J.F. 2004. A regional assessment of the potential stressors causing the decline of the Cherry Point Pacific herring run and alternative management endpoints for the Cherry Point Reserve (Washington, USA). *Hum. Ecol. Risk Assess.,* 10, 271–297.

Landis, W.G., Hart Hayes, E., and Markiewicz, A.M. 2003. Weight of Evidence and Path Analysis Applied to the Identification of Causes of the Cherry Point Pacific Herring Decline. Droscher, T. and Fraser, D.A., Eds., Proceedings of the 2003 Georgia Basin/Puget Sound Research Conference, March 31–April 3, 2003, Vancouver, British Columbia.

Landis, W.G., Matthews, R.A., and Matthews, G.B. 1996. The layered and historical nature of ecological systems and the risk assessment of pesticides, *Environ. Toxicol. Chem.,* 15, 432–440.

Landis, W.G., Luxon, M., and Bodensteiner, L.R. 2000. Design of a Relative Rank Method Regional-Scale Risk Assessment with Confirmational Sampling for the Willamette and McKenzie Rivers, Oregon. *Ninth Symposium on Environmental Toxicology and Risk Assessment: Recent Achievements in Environmental Fate and Transport*, ASTM STP1381 Price, F.T., Brix, K.V., and Lane, N.K., Eds., American Society for Testing and Materials, West Conshohocken, PA, pp. 67–88.

Landis, W.G., Matthews, G.B., Matthews, R.A., and Sergeant, A. 1994. Application of multivariate techniques to endpoint determination, selection and evaluation in ecological risk assessment, *Environ. Toxicol. Chem.*, 12, 1917–1927.

Luxon, M. 2000. Application of the Relative Risk Model for Regional Risk Assessment to the Upper Willamette River and Lower McKenzie River, OR. MS thesis, Western Washington University, Bellingham.

Matthews, R.A., Landis, W.G., and Matthews, G.B. 1996. Community conditioning: an ecological approach to environmental toxicology, *Environ. Toxicol. Chem.*, 15, 597–603.

McLaughlin, J.F. and Landis, W.G. 2000. *Effects of Environmental Contaminants in Spatially Structured Environments. Environmental Contaminants in Terrestrial Vertebrates: Effects on Populations. Communities, and Ecosystems*, Albers, P.H. et al., Eds., Society of Environmental Toxicology and Chemistry, Pensacola, FL, pp. 245–276.

Moraes, R., Landis, W.G., and Molander, S. 2002. Regional risk assessment of a Brazilian rain forest reserve, *Hum. Ecol. Risk Assess.*, 8, 1779–1803.

Obery, A. and Landis, W.G. 2002. Application of the relative risk model for Codorus Creek watershed relative ecological risk assessment: an approach for multiple stressors, *Hum. Ecol. Risk Assess.*, 8, 405–428.

O'Neill, R.V., Hunsaker, C.T, Jones, K.B., Ritters, K.H., Wickham, J.D., Schwartz, P.M., Goodman, I.A., Jackson, B.L., and Baillargeon, W.S. 1997. Monitoring environmental quality at the landscape scale, *BioScience*, 47, 513–519.

Suter, G.W., II. 1990. Environmental risk assessment/environmental hazard assessment: similarities and differences, in *Aquatic Toxicology and Risk Assessment*, 13th vol., ASTM STP1096, Landis, W.G. and van der Schalie, W.H., Eds., American Society for Testing and Materials, Philadelphia, PA, pp. 5–15.

Suter, G.W., II. 1993. *Ecological Risk Assessment*, Lewis Publ., Chelsea, MI, p. 538.

Swartz, R.C., Schults, D.W., Ozretich, R.J., Lamberson, J.O., Cole, F.A., DeWitt, T.H., Redmond, M.S., and Ferraro, S.P. 1995. ΣPAH: a model to predict the toxicity of polynuclear aromatic hydrocarbon mixtures in field-collected sediments, *Environ. Toxicol. Chem.*, 14, 1977–1987.

Thomas, J. 2003. Integration of a Relative Risk Multi-Stressor Risk Assessment with the NCASI Long-Term Receiving Water Studies to Assess Effluent Effects at the Watershed Level, Leaf River, Mississippi, Technical Bulletin No. 867. National Council for Air and Stream Improvement, Research Triangle Park, NC.

Walker, R., Landis, W.G., and Brown, P. 2001. Developing a regional ecological risk assessment: a case study of a Tasmanian agricultural catchment, *Hum. Ecol, Risk Assess.*, 7, 417–439.

Warren-Hicks, W.J. and Moore, D.R.J. 1998. *Uncertainty Analysis in Ecological Risk Assessment*, SETAC Press, Pensacola, FL.

Wiegers, J.K., Feder, H.M., Mortensen, L.S., Shaw, D.G., Wilson, V.J., and Landis, W.G. 1998. A regional multiple stressor rank-based ecological risk assessment for the fjord of Port Valdez, AK, *Hum. Ecol. Risk Assess.*, 4, 1125–1173.

CHAPTER 3

Interactions among Risk Assessors, Decision Makers, and Stakeholders at the Regional Scale: The Importance of Connecting Landscape-Level Endpoints with Management Decisions

P. Bruce Duncan

CONTENTS

1-56670-655-6/05/$0.00+$1.50
© 2005 by CRC Press LLC

INTRODUCTION

This chapter explores the important link between the regional (ecological) risk assessment (RRA) process and the decisions these assessments inform. This link has enabled changes in: (1) how RRAs (and ecological risk assessments [EcoRAs]) are scoped and conducted; (2) the decision-making processes that utilize RRAs and EcoRAs; (3) integration of risk assessment and risk management; and (4) how risk management options are developed. RRA has been defined and compared with EcoRA in Chapter 2. For the purposes of this chapter, both processes are similar enough with respect to their links with management decisions that they are discussed somewhat interchangeably, although distinctions are made where necessary. As illustrated in numerous examples in this chapter, management needs are the foundation on which both EcoRAs and RRAs develop. Important — perhaps difficult or even conflicting — goals and decisions must be identified early in the process. Without identifying, discussing, and resolving these issues, the assessment results will not appear to be useful to managers, and in fact may not be usable for the decisions at hand.

The management decision/risk assessment link has evolved both within the Environmental Protection Agency (EPA) and the broader community of risk assessors and risk managers. This evolution is described by focusing on risk paradigms in general, then looking at EPA-specific guidance as developed and applied within the EPA Superfund program. Next, parallels in the evolution of the management decision/risk assessment link in investigations of ecosystem vulnerability at regional scales are explored. This is done by noting how several programs within EPA's Office of Research and Development (ORD) have come to view the management decision/risk assessment link in the same light as the Superfund program and describing how this link has been approached at several large watersheds. With this backdrop, several RRA case studies are reviewed to help illustrate the state of the practice and to begin to uncover practical methods to improve the interaction between regional risk assessors and risk managers. Some suggestions are offered on how the management decision/risk assessment link is likely to continue evolving in the areas of tribal issues, economics, and adaptive management.

EXAMPLES OF THE MANAGEMENT DECISION/RISK ASSESSMENT LINK IN ECOLOGICAL RISK ASSESSMENT PARADIGMS, FRAMEWORKS, GUIDANCE

The general scientific goals of EcoRAs (including RRAs) are to support decision making in managing ecological resources, create testable hypotheses, and organize and analyze data within a conceptual model. Risk assessments become most effective when they target the decision-making needs and goals of environmental managers.

Because risk assessment and its outcome go beyond data assessment to include decision making, communication, and, often, remedial decisions, many paradigms for conducting ecological assessments now incorporate a close association between the risk assessors, risk managers, and stakeholders (e.g., USEPA 1998c; and see Figure 3.1 from Cirone and Duncan 2000). Eduljee (2000) has recently advocated a pluralistic, inclusive approach, with experts participating alongside other stake-

PLANNING

PROBLEM
FORMULATION

DATA
ACQUISITION

ANALYSIS

RISK
CHARACTERIZATION

COMMUNICATION

RISK MANAGEMENT

Figure 3.1   The importance of risk management as well as how problem formulation feeds into data acquisition and data analysis. (From Cirone, P.A. and Duncan, P.B., *J. Hazardous Mater.*, 78, 1–17, 2000. With permission.) (See color insert following page 178.)

holders in a consensual decision-making process. This interaction is particularly important early in the RRA process when multiple management goals likely exist for a project. In addition, Power and McCarty (2002) have stated strongly that a rigid distinction between assessment and management is not desirable. The distinction, while useful for insulating science from politics, does not allow for the "needed public understanding of risk." Rather, they conclude that a major innovation brought to risk assessment framework design is that analytical activities must be done in a process that (1) allows input from society, (2) emphasizes a deliberative process and stakeholder involvement, and (3) broadens the scope of issues to be included in the problem formulation. This new ground in risk assessment is being embraced, for example, by the formation of a new Center for Conservation Science at the Universities of St. Andrews and Stirling (U.K.). This center focuses on interdisciplinary science in an attempt to ensure the protection of Scotland's natural areas (Kareiva 2001). Conservation policy inevitably involves conflicts among stakeholders, risk analysis, and what Kareiva calls "a treacherous mingling of science and policy."

The importance of management goals has been emphasized in recent EcoRA guidance documents. For example, EPA's guidance from 1998 states, "Ecological risk assessment is a process used to systematically evaluate and organize data, information, assumptions, and uncertainties to help understand and predict the relationship between stressors and ecological effects *in a way that is useful for environmental decision making*" (USEPA 1998c, emphasis added). Released at about the same time, guidance for the Superfund program echoes the importance of this link (USEPA 1997). In the Superfund program, the scientific functions of an EcoRA are to: (1) document whether actual or potential ecological risks exist at a site; (2) identify which contaminants present at a site pose an ecological risk; and (3) generate data to be used

Table 3.1   Steps in the Ecological Risk Assessment Process and Corresponding Decision Points in the Superfund Process (USEPA 1997, Exhibit I-3)

Steps and Scientific/Management Decision Points (SMDPs)	
1. Screening-Level Problem Formulation and Ecological Effects Evaluation	
2. Screening-Level Preliminary Exposure Estimate and Risk Calculation	SMDP (a)
3. Baseline Risk Assessment Problem Formulation	SMDP (b)
4. Study Design and Data Quality Objectives	SMDP (c)
5. Field Verification of Sampling Design	SMDP (d)
6. Site Investigation and Analysis of Exposure and Effects	[SMDP]
7. Risk Characterization	
8. Risk Management	SMDP (e)

Corresponding Decision Points in the Superfund Process

(a) Decision about whether or not a full ecological risk assessment is necessary.

(b) Agreement among the risk assessors, risk manager, and other involved parties on the conceptual model, including assessment endpoints, exposure pathways, and questions or risk hypotheses.

(c) Agreement among the risk assessors and risk manager on the measurement endpoints, study design, and data interpretation and analysis.

(d) Signing approval of the work plan and sampling and analysis plan for the ecological risk assessment.

(e) Signing the Record of Decision.

[SMDP] only if change to the sampling and analysis plan is necessary.

in evaluating cleanup options. In addition to these key points, EPA's Superfund EcoRA guidance very prominently introduced, identified, and described a series of six *scientific-management decision points* (Table 3.1) that show the continual integration and iteration of the link between risk assessment and management decisions (USEPA 1997). Further supporting this linkage, EPA subsequently published the *Ecological Risk Assessment and Risk Management Principles for Superfund Sites* (USEPA 1999). This directive establishes six principles for managers when making ecological risk management decisions, two of which concern risk management decisions: Principle 2 — Coordinate with Federal, Tribal, and State Natural Resource Trustees (to get broad input on the decision and science behind it), and Principle 3 — Use site-specific ecological risk data to support cleanup decisions. This directive also pointedly refers to the Superfund EcoRA process (USEPA 1997), stating that the five key risk assessor/risk manager decision points (Table 3.1) should always be used.

The application of the guidance has been successful at many Superfund sites. One example from EPA, Region 10, illustrates this especially well. At the Eagle River Flats artillery training site for the U.S. Army (Fort Richardson, AK) (USACE 2004), waterfowl were dying from effects of ingesting pellets remaining from the incomplete combustion of white phosphorus (WP) used in smoke rounds. Unacceptable risk was obvious. The focus of the assessment efforts was directed to the main question of what remedial action should be taken and where. The EPA, the U.S. Army, the U.S. Fish and Wildlife Service, the Alaska Department of Fish and Game, the Alaska Department of Environmental Conservation, and the U.S. Army Toxic and Hazardous Materials Agency (now known as the U.S. Army Environment Center) reached consensus on an approach to characterize risk. Data on WP concentrations;

crater density; waterfowl habitat, use, and mortality; and physical system dynamics (e.g., erosion) were drawn together into a geographical information system to prioritize areas for remediation. Most importantly to managers, scientific/management agreement was reached on the remedial action objectives, namely: the 5-year goal is a 50% reduction in waterfowl mortality, based on 1996 mortality data, and the 20-year goal is a reduction in mortality to 1% of the total population. Goals and endpoints are measured as a mortality percentage of radio-collared mallards captured and released on site. It was recognized that this tool eventually would not be sensitive enough to statistically determine increasingly lower mortality rates, so the Record of Decision, Section 7.2 (USEPA 1998b) states that at the 5-year review " ... the telemetry results, interpretation methods, and remedial action objectives will also be re-evaluated." A 5-year review, conducted in 2003 before active remediation had ceased, noted that the short-term objective had been met, the long-term objective had not, and that recovery trends should continue to be evaluated (USEPA 2003b).The first full 5-year term review period following remediation will occur in 2008. It is likely that it will include a discussion of all the lines of evidence relating to achievement of remedial success, even if mortality estimated from telemetry is not a practical method (Bill Adams, EPA Project Manager, 2003, personal communication).

EVOLUTION OF THE MANAGEMENT DECISION/RISK ASSESSMENT LINK IN REGIONAL-SCALE ASSESSMENTS

Risk assessors and managers involved in regional-scale assessments are realizing how important it is to include the link between management decisions and risk assessment in evaluating relative risks or ecosystem vulnerability. For example, a prototype tool comparing small watershed units could only be developed once management goals were defined. These goals included potential actions such as protection, restoration, monitoring, or collecting additional information. The prototype was developed to illustrate how the proximity of stressors to resources (organisms, habitats, etc.) within each watershed unit could be used to identify areas for each potential action (USEPA 1998a). A recent review of comparative risk projects (Feldman et al. 1999) showed that "while many priority-setting projects have successfully identified environmental problems and characterized and ranked their risks, few have developed risk-management strategies." Indeed, the authors of that report go on to say that "... no project that we evaluated has, as yet, documented achievement of a system for developing and implementing environmental priorities in order to mitigate their most significant environmental problems." Since then, the relative risk model, discussed in Chapter 10 in detail, has emerged as a successful tool when conducting RRAs and ranking risks. This tool has been shown to work well at several geographic scales, across a variety of aquatic landscapes (Landis et al. 2000; Walker et al. 2001; Moraes et al. 2002; Obery and Landis 2002; Landis et al. 2004).

Investigators of ecosystem vulnerability at regional scales have also noted the importance of linking risk assessment and management decisions to improve EcoRA and management practices. To accomplish the improvements, community management has emerged as an issue that needs to be incorporated in regional-scale assessments. A symposium on Modeling and Measuring the Vulnerability of Ecosystems at Regional Scales for Use in Ecological Risk Assessment and Risk Management (August 17–20, 1998, Seattle, WA) was organized and cosponsored by the EPA, the Society of Environmental Toxicology and Chemistry (SETAC), and the American Society of Testing Materials (ASTM) Committee E47 on Biological Effects and Chemical Fate, to discuss problems and explore potential solutions to integration of data and models across scales of time and space, levels of organization, and multiple stressors. It provided an important venue for fostering communication, especially between the social sciences and the natural sciences. A workshop subsequent to the symposium identified community management as a major issue and presented these findings (Williams et al. 2000):

- While humans define the values by which we determine ecosystem health goals, we must recognize that ecosystem health values go beyond solely human values and that humans are both stressors and receptors in a community and, with their biases, define the values by which we determine ecosystem use goals (see also Cirone and Duncan 2000).
- They must educate (and be educated) as part of community involvement using jargon-free dialogue and must learn how to articulate their concerns.
- The local human community is good at identifying local scale and some broader impacts, and should be involved in discussions of the issues of concern.
- Community management must recognize that community extends as far as the most distant stressor source.
- The community must realize that action may need to occur at a scale different than the problem recognition.

The EPA has had a long, ongoing involvement with regional assessments and has identified the value of partnering with decision makers at more local levels so that tools developed at large scales can be effectively applied at finer scales. For 10 years, EPA's ORD has been involved with the Mid-Atlantic Integrated Assessment (MAIA), a federal, state, and local partnership led by EPA, Region 3. MAIA is a comprehensive regional assessment that focuses on understanding ecosystem processes on a variety of scales. Based on this involvement, the managers and scientists working on MAIA recently identified five steps to improving environmental decision making (Table 3.2) (Smith et al. 2000). Interestingly, it is only at the fourth step that policy is addressed. The earlier steps have produced "the most complete set of data on regional environmental condition and trends in the United States. As part of the MAIA, EPA's EMAP (Environmental Monitoring and Assessment Program), the regional office, and state and local partners have produced environmental report cards on the health of highland streams, estuaries, and a landscape assessment" (Smith et al. 2000). The people involved with MAIA have recognized that making their science useful to stakeholder and management concerns in the regions in which the assessments occur requires a focused effort.

In response, ORD has developed the Regional Vulnerability Assessment Program (ReVA) to be involved with policy decisions and prioritization of stressors. However, the scientists assume that assessment endpoints (i.e., concerns related to the ecosystems)

Table 3.2   Iterative Steps to Improving Environmental Decision Making from EPA's Involvement in the Mid-Atlantic Integrated Assessment (MAIA)

1. Monitoring to establish status and trends
2. Association analyses to suggest probable cause where degradation is observed
3. Prioritization of the role of individual stressors as they affect cumulative impacts and risk of future environmental degradation
4. Analysis of the trade-offs associated with future policy decisions
5. Development of strategies to restore areas and reduce risk

Note: Environmental Monitoring and Assessment Program (EMAP) is developing approaches to address steps 1 and 2; Regional Vulnerability Assessment Program (ReVA) is developing approaches to address steps 3 and 4. Approaches to step 5 will be addressed in a new research program that is under development (Smith et al. 2000).

have already been identified by regional stakeholders and that data for these endpoints are available. Therefore, although scientists in the ReVA program are developing approaches to address steps three and four (Table 3.2) that deal with an analysis of the trade-offs associated with future policy decisions, their main goal is to link assessment endpoints with regional stressors. Despite focusing on the regional scale, the scientists recognize that effective environmental decision making often occurs "at scales below the regional, i.e., local, community, and watershed" and so they plan to partner with clients at the state and local level to develop finer-scale applications of the regional assessment information (Smith et al. 2000). The thrust of ReVA is clearly to develop tools and transfer these to decision makers operating at various scales. It is encouraging that the program will partner with stakeholders to develop the applications. The partnering will be done through "close interactions with EPA Region 3 and through periodic review by a diverse group of regional stakeholders" with the hope that close interaction with regional decision makers as the assessment methodology is developed will ensure that "the appropriate questions are posed and that research results are more widely disseminated than has traditionally been done" (USEPA 2003a).

The ReVA includes the State of Pennsylvania's Department of Environmental Protection (PADEP) as one of its local partners. The PADEP has implemented an Environmental Futures Planning Process which provides a good example of the recognized need for a partnership with a broad group of stakeholders at a regional scale (PADEP 2003). Begun in 2001, this is a three-step process, repeated every year, that answers: What are the conditions in the environment, and why? What are the targets to improve those conditions? And, what are the detailed plans to meet those targets? Input has been received by the PADEP from a variety of stakeholders across the state, including planning objectives from 34 watershed teams, as well as central office bureaus. The plans let the public see how PADEP is addressing environmental issues. Partners help create and implement some of the plans and those involved are able to see how their efforts produce real environmental outcomes. The partners are broad-based and include: advisory boards on agriculture, cleanup standards, mining and reclamation, oil and gas, and technical assistance; advisory committees on air quality, bituminous mine safety, Chesapeake Bay waters, small business compliance, solid waste, and water resources; Citizen's Advisory Councils; and the Pennsylvania Association of Conservation Districts. Action plans were developed by mid-2002 and presented at over 75 public meetings. It is not clear whether momentum has continued, actions and feedback have been incorporated into the process, or this large upfront investment has paid off.

Upon examining some projects that have tackled watershed-scale management issues, there appears to be a transition occurring from presenting management options *to* stakeholders to developing management options *with* stakeholders. For example, implementing a comprehensive plan to protect the Big Darby, OH watershed from stormwater impacts has been complex, due to the many political entities involved (Jones and Gordon 2000). To engage public officials and other interested citizens, a series of meetings (cosponsored by the local Soil and Water Conservation Districts and Ohio State University Extension) were held in parts of the watershed to determine their reaction to three options for managing storm water. The meetings began with a presentation illustrating the nature of stormwater problems and potential management options. Presentations were also made at a regional planning coordinating meeting, after which informal discussion was encouraged to again determine stakeholder reaction. Broad-based consensual support existed from around the watershed concerning the importance of preserving the quality of the Big Darby Creek watershed. "But although a great deal of agreement existed on where to go, much less agreement occurred on how to get there from here" (Jones and Gordon 2000).

Like the Big Darby, the Middle Snake River, ID watershed assessment illustrates the evolving role for risk assessment applied to multiple stressors and resources, particularly the outcome of using a consensus-building method to reach solutions. In addition to advancing the science of watershed-level assessments, a major reason for the Middle Snake River assessment was to "ensure that the public and special-interest users, government agencies, and scientists understand the ecological damage and that they develop a sense of partnership in reaching solutions for the recovery and protection of the Middle Snake River ecosystem" (USEPA 2002). The assessment drew from the experience of the Middle Snake River Watershed Council, which had evolved from interested groups working together since the mid-1980s (USEPA 2002). The management goals for the Middle Snake are largely driven by state and federal legislation as well as county landuse plans and include "attainment of water quality standards, establishment of total maximum daily loads for major pollutants, water for hydropower, recreation, and irrigation, recovery of endangered species, and sustained economic well being" (USEPA 2002).

A good example of including stakeholders early is provided by Gentile et al. (2001) concerning the complex environmental issues in the Florida Everglades, where they describe the application of the problem formulation phase of the EcoRA process at regional scales. The three main steps in their problem formulation were (1) an initial planning step that integrates scientific, management, stakeholder, and public preferences and values into a clear statement of goals and objectives for the study; (2) the identification and selection of a suite of ecological endpoints that capture the health of the system; and (3) the development of a conceptual model that describes, qualitatively or quantitatively, the potential causal relationships among human activities in the landscape, system drivers, stressors, and ecological systems. The importance of identifying management needs/decisions at the very beginning of an EcoRA or RRA is clear, and as Gentile et al. (2001) state, "The conceptual model, therefore, is the single most important product of the problem formulation exercise and a critical component of the risk assessment, management, and recovery process."

STATE OF THE PRACTICE IN LINKING MANAGEMENT DECISIONS AND RISK ASSESSMENT IN RRAS

Five RRA projects are discussed chronologically in terms of their identified management goals. To understand the importance of management needs on the direction of the RRA, one would ideally like to have the following information: How were management/policy goals identified? Were competing goals identified, and how were they dealt with? When and how did the risk assessors interact with managers? What were key problems and successes? and, What were the key lessons learned? Apart from the first question, these aspects of risk assessment are generally not reported.

Valdez

This assessment (Wiegers et al. 1998) applied the relative risk model (RRM) that is used to numerically rank and sum risks by stressor, area, or both (Landis and Wiegers 1997). This RRA also described how stakeholders, managers, and assessors develop assessment endpoints. The geographic area was Port Valdez, Alaska. Ecosystems of concern included bays, shorelines, shallow subtidal areas, and basins. The overarching purpose for the RRA was to assess risk to (1) water and sediment quality in Port Valdez; (2) finfish and shellfish populations used by fishermen; and (3) wildlife populations (fish, birds, and mammals) utilizing the Port. Three public meetings were held in the City of Valdez in October 1995 to aid in the formulation of assessment endpoints relevant to the Port. Following a brief introduction to the risk assessment process, the public was asked about their concerns for the Port Valdez environment. Responses were sorted into two general categories: (1) stressors and sources of concern in the Port and (2) populations or attributes of the Port that people wished to protect. Interviews in the community supplemented the public meetings. Participants included the city planning department, the Alaska Department of Environmental Conservation, the U.S. Coast Guard, as well as local industry managers.

Tasmania

This was a preliminary RRA (Walker et al. 2001), focused on obtaining stakeholder input. The geographic area was Mountain River (Huon Valley) in Tasmania, an area of approximately 190 km². The stream system is influenced by horticulture and agriculture. The regulatory setting was well characterized — and more emphasis given on integration with stakeholders than in the Valdez assessment. For example, the goals of the local community were used to develop assessment endpoints. A community forum held in 1998 identified the following issues: improve water quality (particularly decrease *E. coli* counts), maintain/establish water of drinkable and irrigatable quality, maintain habitats for aquatic animals, water in suitable volumes to sustain agriculture, catchment quality for town water supply, water for swimming, water for trout fishing, maintain and improve beauty of the river, and maintain seasonal nature of the river. At a 1999 catchment community forum, locals created an image of their preferred catchment having the following characteristics: clean water that is safe for drinking and swimming, sustainable landuse practices, optimum stream flow, natural vegetation along the riverbanks, an active and responsible community, and an attractive setting for picnics. This forum was not well attended by local farmers, which led to one of the lessons learned. "In a preliminary risk assessment such as this, perhaps the most important function is collation of information about the region, and focus on what stakeholders want for the region ... It is vital that assessment endpoints be determined with a conscientious and intelligent effort to represent the values of the entire community."

Codorus Creek

This is a good example (Obery and Landis 2002) of linking assessment and management issues in problem formulation, covering stakeholder input and the regulatory setting. The geographic area was York County, PA, where the Codorus Creek Watershed (CCW) drains an area of 719 km². Ecosystems of concern included surface waters, perennial streams, and the creek itself (which ranges from 1 to 36 m wide). Regions were identified by land use (industrial, agriculture, etc.). The overarching purpose for the RRA was to identify risk to stakeholder values and potential mitigation, as well as test the applicability of the RRM beyond the Port Valdez case. The regulatory setting was well described (e.g., impaired waters requiring total maximum daily load estimated; fishery uses). Management goals (used to develop assessment endpoints) were identified at a CCW Association meeting, which included representatives from various stakeholder groups such as the PADEP, local industries, Trout Unlimited, and local citizens. The goal was to provide individual stressor and habitat ranks, areas to be protected, areas of high stress, and areas where additional information should be collected.

Brazil

This assessment (Moraes et al. 2002) is a prototype that argues for inclusion of societal economic considerations. The *Parque Estadual Turístico Alto do Ribeira* (PETAR) is a natural reserve in southeastern Brazil. Two river ecosystems components, epigean (surface) and hypogean (subterranean), were evaluated. The overarching purpose for the RRA was to determine which of the different regions in three catchment areas were more likely to be impacted by different forms of land use inside and near PETAR. In addition, the study compared risks from different stressors to the aquatic fauna of the park. It was unclear how assessment endpoints were selected and it appeared that this RRA was intended to be a case study to show how the RRA model can be used elsewhere in Brazil and other tropical areas in developing countries. Only potential uses of the results were discussed. The positive merits of RRA were clearly indicated: low demand for input data, transparent assessment models, low cost, and ease of use in risk communication and prioritization. Social conditions were discussed. The authors argue that, since PETAR is located in one of the poorest areas in the country, management actions must "balance the benefits of preservation against the costs of limiting economic growth of the region by inclusion of agriculturally valuable land in the reserve."

Cherry Point Reserve

This assessment (Landis et al. 2004) illustrates the importance of having clear management goals, and how understanding the goals led to a change in the management approach to the problem at hand. When Western Washington University began to assist the Washington Department of Natural Resources (WDNR) with its investigation of the herring decline at the Cherry Point Reserve, the investigation had branched out to a number of separate studies of potential causes, results from which were discussed at a local chapter meeting of SETAC. The issue of the decline appeared to be an excellent candidate for a RRA; there were multiple potential stressors and a large geographic range (encompassing the life history of the Cherry Point herring stock). The basic steps in RRA were laid out: (a) list management goals for the region, (b) make a map, (c) break the map into regions, (d) develop a conceptual model, (e) decide on ranking scheme, (f) calculate the relative risks, (g) evaluate the uncertainty, (h) generate testable hypotheses, and (i) test the hypotheses, and (j) communicate the results.

From the outset, WDNR was asked to provide its management goals relative to the Cherry Point herring. It turned out that the WDNR had authority for multiple management decisions in that and nearby geographic areas (Table 3.3). These pending decisions (about stressors) indicated that WDNR risk assessments are geographically based, across wide areas, with multiple stressors (and multiple resources). It was also clear that mitigation, restoration, and protection are as important as leasing and permitting to WDNR and that WDNR could apply RRA to leases, easements, and rights of way; aquatic reserve management decisions; restoration projects; public access decisions; and actions to prevent "takings" under the Endangered Species Act. These considerations helped move from a herring-centric framework to focus on WDNR management needs and affirm the goal of the risk assessment to aid the decision-making process for the management of the natural resources within the Cherry Point region. This in turn led to a broader consideration that ecological management of the Cherry Point Reserve requires information specific to Cherry Point and the organisms living there, and identification of several candidate indicators species (other than Pacific herring) that are culturally or commercially important, utilize habitat with a high probability of exposure to contaminants and other stressors, are year-round residents, and are connected ecologically to the Pacific herring (Hart Hayes and Landis 2004). The RRA concluded that, because adult herring range widely and are influenced by large-scale stressors outside Cherry Point, herring are a poor measure of the ecological status of the Cherry Point area.

Table 3.3   Washington Department of Natural Resources (WDNR) Decisions Relative to Ecological Risk Assessment

The Kinds of Decisions to which WDNR May Apply Risk Assessment

Leases, easements, rights of way
Aquatic reserve management decisions
Restoration projects
Public access decisions
Actions to prevent "takings" under the Endangered Species Act

Primary Area: Cherry Point Reach — Point Whitehorn to Sandy Point

Decisions Pending
Alcoa/Intalco pier and outfall lease renewal (exp. 1998)
Alcoa/Intalco stormwater outfall (perpetual easement)
Lummi Indian Business Council waste outfall, Neptune Beach (exp. 1998)
PIT proposal for dock expansion (decision needed in next 18 to 24 months)
Williams Pipeline request to cross reserve with pipeline (pending)
Whatcom Co. waste outfall (2005)
Tosco pier and outfall lease renewal (2008)
BP/ARCO lease compliance monitoring (2029)
How to deal with abandoned sand and gravel structures at Gulf Road
How to deal with outfalls not under WDNR lease (Unick Road, etc.)
Aquatic reserve management — what actions are compatible with goals of reserve?
Will a given action help avoid "takings" under the Endangered Species Act?
Evaluate restoration and mitigation projects (such as sediment cleanup, remediation, or habitat improvement projects) for how well they further goals for region
How big an impact do recreational uses (like clam digging) have on the area?

Secondary Area: Bellingham Bay Aquatic Landscape Planning Area

Point Whitehorn south to Lummi Bay, Bellingham Bay, and Chuckanut Bay (Chuckanut, Portage, Eliza, and Lummi Islands)

Decisions Pending
How to mesh Cherry Point decisions with this larger landscape
Managing specific structures, activities, or geographic features that have an impact on conditions at Cherry Point, or, conversely, Cherry Point items that have an impact on this broader area
Bellingham Bay Pilot Project (sediment cleanup)
Other leases and activities in this area that have an impact

Tertiary Area: Northwest Straits Planning Area

Aquatic Lands in Skagit, Whatcom, Island, San Juan, Snohomish, Clallam, and Jefferson Counties
No decisions in this area have currently been linked specifically to Cherry Point, but there are 74 major use leases and other activities that we manage in this area.

Source: From Landis et al., *Hum. Ecol. Risk Assess.*, 10, 271–297, 2004. With permission.

DISCUSSION AND CONCLUSIONS — IMPROVING THE LINK AND LIKELY NEXT STEPS IN ITS EVOLUTION

The examples discussed above clearly indicate that scoping in EcoRAs and RRAs has changed (recommended by USEPA [1998c]; Eduljee [2000]; and Moore [2001]) to include stakeholder values (e.g., the Florida Everglades problem formulation, Gentile et al. 2001; and the Codorus Creek RRA, Obery and Landis 2002) as well as broader interaction with the community as called for by Williams et al. (2000) and put into practice, for example, by Walker et al., (2001), Obery and Landis (2002), and PADEP (2003). This has been a natural evolution fostered by the need to develop assessment endpoints in risk assessments and, to some extent, the need to consider cultural endpoints and tribal perspectives that are, by nature, more inclusive than the now-outdated traditional risk assessment/risk management dichotomy. This change, the linking of stakeholders, managers, and assessors in risk assessment, has been characterized by Power and McCarty (2002) as a major innovation. As these changes in the analytic assessment process become more integrated with the decision-making process they will continue to change the latter. These linkages allow risk management options to be productively developed and focused (i.e., Cherry Point RRA) and showcase the strengths of RRA to inform risk management decisions.

Based on the examples presented in this chapter, there is a strong need to develop guidelines and examples that codify collective experiences in decision making and stakeholder involvement. There is ample guidance saying that stakeholders and risk assessors should be incorporated into the risk management process, and an entire field of risk communication. Even guidance on risk characterization (one of the final steps of risk assessment) for watershed-scale risk assessments (Serveiss et al. 2000) calls for (1) "... regular consultation with risk managers and stakeholders ... throughout the process" (Serveiss 2002), (2) realization that such ongoing involvement could change the direction of the assessment (which it has; see Hart Hayes and Landis [2004]), and (3) adding regular consultation to USEPA's (1998c) Guidelines.

What is sorely lacking at present is a summary of practical experience on what actually allows breakthroughs to be reached when the "treacherous mingling of science and policy" (Kareiva 2001) is taking place. Agreements need to be reached on management goals, assessment endpoints, remedial actions, etc. For example, when beginning the Cherry Point RRA, half-day training was provided to WDNR managers on basic RRA methodology and a full day of technical training was provided to agency staff. Through RRA a major breakthrough occurred when WDNR was asked for and provided a detailed list of management goals for the region where the herring stock was in decline. In the simplest terms, the key in all assessments is to seek upfront agreements on project scope and use of the results. Unfortunately, details on how this actually occurs are generally never reported. Clearly, it is an opportune time to develop guidelines for incorporating stakeholders (i.e., interested parties) and assessors into risk management.

It is likely that there is much to be learned from decision-making processes outside of the current RRA paradigm. For example, the U.S. Geological Survey has a program composed of social scientists conducting research on the role technical clarity has in reaching successful multiparty agreements concerning science-based questions (Lamb et al. 2001). Similarly, other processes that embrace stakeholder involvement may have approaches and frameworks useful to developing the important nexus between decision makers, risk assessors, and the affected community.

This nexus seems poised to embrace other significant assessment/management issues such as tribal issues, economics, and adaptive management. For example,

Williams et al. (2000) of the Department of the Environment of the Mohawk Council of Akwesasne has developed a way of looking at a tribal community (Naturalized Knowledge Systems) to acquire knowledge about the place where members live. This approach uses new tools while still maintaining traditional concepts. It is represented by six basic principles: the Earth is Our Mother, cooperation is the way to survive, knowledge is powerful only if it is shared, the spiritual world is not distant from Earth, responsibility is the best practice, and everything is connected to everything. The successful implementation of the Naturalized Knowledge System relies on a balance of respect, equity, and empowerment. It will be very interesting to see how this aspect of community values will be incorporated into risk assessment/management frameworks and then implemented.

Like tribal issues, economic issues enter risk assessments whether transparently included or not. For example, Power and McCarty (2002) point out that due to limited resources for risk management, and improved abilities to identify and assess risk, it has become increasingly necessary to ensure that risk reductions are achieved at reasonable cost. Concomitantly, changes in attitudes toward the use of socioeconomic information, risk characterization, and uncertainty analysis are also evident. The trend has favored greater emphasis on the inclusion of economic analysis in decision making through the direct estimation of benefit and cost. Despite recognizing the role economics can play, frameworks that include it (Power and McCarty 2002) are careful to note that it should not be the overriding determinant in decision making.

Although assessments have almost always had iterative elements and feedback loops, the push is now for a clear adaptive management strategy. Feldman et al. (1999) in their evaluation of comparative risk projects note that it is difficult to know if and when a system for developing and implementing environmental priorities to mitigate significant environmental problems is successful unless projects establish mechanisms for evaluating their results. In his emphasis on integration in risk assessment, Moore (2001) points out that in addition to societal and political buy-in to the assessment and decision-making process, the assessment must (1) be able to consider a wide range of stressors and potential risk management options, (2) become focused, and (3) adopt an adaptive management strategy. In the adaptive management strategy risk management actions are undertaken, system response intensively observed and assessed, and revised management actions taken as appropriate (Moore 2001). In response to this type of compelling logic, EPA's ReVA program now includes a new research program to develop strategies to restore areas and reduce risk (Smith et al. 2000; Table 3.2 this volume). This program is one that is expected to evolve over time and, as informed decision making is implemented, the integrated assessment approach should also provide feedback to ORD on the success of its research and development activities in terms of actual improvement in environmental quality (USEPA 2003a). Certainly, feedback becomes critical when it "is important that progress be apparent," even in the face of deadlines, limited resources, and the continued habitat decline (USEPA 2002).

In summary, the conclusions based on examining the link between risk assessment and management decisions resonate with common sense. Lamb et al. (2001) have demonstrated how technical clarity is critical to negotiations of science-based issues. Similarly, when a risk assessor has access to clear management goals and understands the decisions at hand, then a successful risk assessment can be designed. This understanding is what focuses and streamlines the EcoRA; it is one of the first steps even though the final decisions are made at the last step. The benefits begin at the planning stage of the EcoRA and cascade throughout the risk assessment, extending to risk managers and stakeholders in the larger community where the effects of the decisions ultimately reside.

REFERENCES

Cirone, P.A. and Duncan, P.B. 2000. Integrating human health and ecological concerns in risk assessments, *J. Hazardous Mater.*, 78, 1–17.

Eduljee, G.H. 2000. Trends in risk assessment and risk management, *Sci. Total Environ.*, 249, 13–23.

Feldman, D.L., Hanahan, R.A., and Perhac, R. 1999. Environmental priority-setting through comparative risk assessment, *Environ. Manage.*, 23, 483–493.

Gentile, J.H., Harwell, M.A., Cropper, W., Jr., et al. 2001. Ecological conceptual models: a framework and case study on ecosystem management for South Florida sustainability, *Sci. Total Environ.*, 274, 231–253.

Hart Hayes, E. and Landis, W.G. 2004. Regional ecological risk assessment of a nearshore marine environment: Cherry Point, WA, *Hum. Ecol. Risk Assess.*, 10, 299–325.

Jones, A.L. and Gordon, S.I. 2000. From plan to practice: Implementing watershed-based strategies into local, state, and federal policy, *Environ. Toxicol. Chem.*, 19, 1136–1142.

Kareiva, P. 2001. Risk assessment and stakeholder-based decision making, *Trends Ecol. Evol.*, 16, 605–606.

Lamb, B.L., Burkardt, N., and Taylor, J.G. 2001. The importance of defining technical issues in interagency environmental negotiations, *Publ. Works Manage. Policy*, 5, 220–232.

Landis, W.G., Markiewicz, A., Thomas, J. et al. 2000. *Regional Risk Assessment for the Cherry Point Herring Stock*, Institute of Environmental Toxicology and Chemistry, Huxley College, Western Washington University, Bellingham, WA.

Landis, W.G., Duncan, P.B., Hart Hayes, E., Markiewicz, A.J., and Thomas, J.F. 2004. A regional assessment of the potential stressors causing the decline of the Cherry Point Pacific herring run and alternative management endpoints for the Cherry Point Reserve (Washington, USA), *Hum. Ecol. Risk Assess.*, 10, 271–297.

Landis, W.G. and Wiegers, J.K. 1997. Design considerations and a suggested approach for regional and comparative ecological risk assessment, *Hum. Ecol. Risk Assess.*, 3, 287–297.

Moore, D.R.J. 2001. The Anna Karenina principle applied to ecological risk assessments of multiple stressors, *Hum. Ecol. Risk Assess.*, 7, 231–237.

Moraes, R., Landis, W.G., and Molander, S. 2002. Regional risk assessment of a Brazilian rain forest, *Human Ecol. Risk Assess.*, 8, 1779–1803.

Obery, A.M. and Landis, W.G. 2002. A regional multiple stressor risk assessment of the Codorus Creek watershed applying the relative risk model, *Hum. Ecol. Risk Assess.*, 8, 405–428

PADEP (Pennsylvania Department of Environmental Protection). 2003. Environmental Futures Planning Process — Setting Priorities for the 21st Century. http://www.dep. state.pa.us/hosting/efp2.html (accessed December 29, 2003).

Power, M. and McCarty, L.S. 2002. Trends in the development of ecological risk assessment and management frameworks, *Hum. Ecol. Risk Assess.*, 8, 7–18.

Serveiss, V.B. 2002. Applying ecological risk principles to watershed assessment and management, *Environ. Manage.*, 29, 145–154.

Serveiss, V.B., Cox, J.P., Moses, J., et al. 2000. Workshop Report on Characterizing Ecological
     Risk at the Watershed Scale (Arlington, VA). EPA/600/R-99/111. Office of Research
     and Development, National Center for Environmental Assessment, Washington, D.C.,
     http://www.cfpub.epa.gov/ncea/cfm/recordisplay.cfm?deid=23760.html.
Smith, E.R., O'Neill, R.V., Wickham, J.D., Jones, K.B., Jackson, L., Kilaru, J.V., and Reuter,
     R. 2000. The U.S. EPA's Regional Vulnerability Assessment Program: A Research
     Strategy for 2001–2006. U.S. Environmental Protection Agency, Office of Research
     and Development, Research Triangle Park, NC, http://www.epa.gov/nerlesd1/land-
     sci/ReVA/reva-strategy.pdf.
USACE (U.S. Army Corps of Engineers). 2004. Eagle River Flats: History, Bibliography,
     Remediation Data, Photographs, Web Camera, Maps, Monitoring, Ecology.
     http://www.crrel.usace.army.mil/erf (accessed January 2, 2004).
USEPA (U.S. Environmental Protection Agency). 1997. Ecological Risk Assessment Guid-
     ance for Superfund: Process for Designing and Conducting Ecological Risk Assess-
     ments, Interim Final. EPA 540-R-97-006. EPA, Washington, D.C., http://epa.gov/
     superfund/programs/risk/ecorisk/ecorisk.htm.
USEPA (U.S. Environmental Protection Agency). 1998a. Comparative Ecological Risk: Using
     the Proximity of Potential and Actual Stressors to Resources as a Tool to Screen
     Geographical Areas for Management Decisions. Comparative Geographical Risk
     Assessment in EPA Region 10, Development of a Prototype. Office of Environmental
     Assessment, Seattle, WA.
USEPA (U.S. Environmental Protection Agency). 1998b. EPA Superfund Record of Decision:
     Fort Richardson (USARMY); EPA ID: AK6214522157; Operable Unit 03 Anchorage,
     AK. EPA/ROD/R10-98/182. EPA, Region 10, Seattle, WA. http://www.epa.gov/
     superfund/sites/rods/fulltext/r1098182.pdf.
USEPA (U.S. Environmental Protection Agency). 1998c. Guidelines for Ecological Risk
     Assessment. EPA/630/R095/002F. Risk Assessment Forum, Washington, D.C., http://
     cfpub.epa.gov/ncea/cfm/recorddisplay.cfm?deid=12460.
USEPA (U.S. Environmental Protection Agency). 1999. Ecological Risk Assessment and Risk
     Management Principles for Superfund Sites. OSWER Directive 9285.7-28 P. EPA,
     Washington, D.C. http://www.epa.gov/superfund/programs/risk/final10-7.pdf.
USEPA (U.S. Environmental Protection Agency). 2002. Ecological Risk Assessment for the
     Middle Snake River, ID. EPA/600/R-01/017. National Center for Environmental
     Assessment, Washington, D.C., http://cfpub.epa.gov/ncea/cfm/recordisplay.cfm?
     deid=29097 and http://www.epa.gov/ncea.
USEPA (U.S. Environmental Protection Agency). 2003a. Regional Vulnerability Assessment
     Program: Approach. Available at http://www.epa.gov/reva/approach.html (accessed
     December 29, 2003).
USEPA (U.S. Environmental Protection Agency). 2003b. ROD-5 year review. EPA, Region
     10, Seattle, WA. http://www.yosemite.epa.gov/R10/CLEANUP.NSF/9f3c21896330b
     4898825687b007a0f33/34319334228615b088256516006cdccd/$FILE/Fort%20Rich
     %205%20Year.pdf.
Walker, R., Landis, W.G., and Brown, P. 2001. Developing a regional ecological risk assess-
     ment: A case study of a Tasmanian agricultural catchment, *Hum. Ecol. Risk Assess.*,
     7, 417–439.
Wiegers, J.K., Feder, H.M., Mortensen, L.S. et al. 1998. A regional multiple-stressor rank-
     based ecological risk assessment for the fjord of Port Valdez, Alaska, *Hum. Ecol.
     Risk Assess.*, 4, 1125–1173.
Williams, R., Llewellyn, R., and Kapustka, L.A. 2000. Ecosystem vulnerability: A complex
     interface with technical components, *Environ. Toxicol. Chem.*, 19, 1055–1058.

CHAPTER 4

Application of the Relative Risk Model to the Fjord of Port Valdez, Alaska

Janice K. Wiegers and Wayne G. Landis

CONTENTS

1-56670-655-6/04/$0.00+$1.50
© 2004 by CRC Press LLC

INTRODUCTION

While the field of ecological risk assessment (EcoRA) is moving toward more systems-based, as well as more realistic, assessments, there is yet little guidance on how to integrate the complex relationships that can exist within environments affected by natural and anthropogenic stresses. Researchers are beginning to call for and to develop qualitative modeling procedures that will help to integrate these components (Harris et al. 1994; Dambacher, Li, and Rossignol 2003). Qualitative models are capable of larger-scale perspectives through which the more specific and quantitative models can be understood. Qualitative models can be used as a framework in which to sort out complex sets of relationships, while the more detailed and quantitative studies usually assess only a couple of variables at a time. In 1997, we developed a relative risk model (RRM) to provide such a framework for Port Valdez, Alaska (Wiegers et al. 1998).

This project was instigated by local concern that activities associated with the Trans Alaska Pipeline were negatively affecting the ecology of the Port. The Regional Citizen's Advisory Committee (RCAC), which provides citizen oversight for pipeline activities, funded the project. To address the varied concerns of the public and the RCAC, we found it necessary to modify the standard risk assessment approach. Modifications resulted in the first application of the RRM, and attained a regional perspective from which we were able to evaluate the risk associated with pipeline activities within the greater context of all activities within the Port. The regional approach requires study of ecological systems at a larger scale as well as consideration of various physical, chemical, and biological stressors that could affect the environment, but are usually not considered within the same assessment. To achieve a more balanced evaluation of the threat to marine populations and communities, we based our assessment on prototypical habitats and anthropogenic sources of stressors. This model considers not only the direct stressors and the organisms affected by these stressors, but also the sources producing these stressors and the habitats on which the organisms depend. A detailed analysis of the risk assessment for Port Valdez is available in Wiegers et al. (1997).

PROJECT BACKGROUND

The primary activity driving public concern for the Port waters was the discharge of up to 21 million gallons of treated ballast water. Ballast water is stored in the cargo holds of oil tankers and transported to the marine terminus of the pipeline located on the south shore of the Port. The terminus is known as the Valdez Marine

Terminal. The ballast water, which is contaminated with crude oil residuals from the ships' previous cargo, is discharged to the ballast water treatment plant (BWTP) and treated through processes of settling, dissolved air flotation, and biological degradation. The effluent is then released into the Port under a National Pollution Discharge Elimination System (NPDES) permit. Low levels of hydrocarbons are known to be present in the effluent.

Despite efforts by the facility to meet regulatory standards and stay in compliance, the large volumes of treated water discharged into the Port create uncertainty in the minds of stakeholders regarding the degree to which hydrocarbons are accumulating in and impacting the marine environment. At the beginning of this project, an EcoRA was planned to evaluate the effect of the effluent chemistry on the Port ecology. The EcoRA was to be based on available data, including effluent testing results, and Port-wide environmental monitoring analyses. Early in the process, several facts emerged suggesting that traditional EcoRA would not provide the best understanding of the potential harm to this environment:

- The influent composition was controlled through best management practices in place for the treatment plant and tanker operations. For instance, only cleaning agents approved by the U.S. Environmental Protection Agency (EPA) could be used on tankers — limiting the potential for chlorinated solvents to be present in the effluent. In addition, the RCAC was monitoring ballast water in tanker holds for the presence of hazardous materials. Due to these controls, the general composition of the effluent was fairly well defined.
- For several years, the effluent had generally met the NPDES requirements for hydrocarbons, including benzene, toluene, ethylbenzene, xylenes (BTEX), naphthalene, and other polycyclic aromatic hydrocarbons (PAHs). Prior exceedences of the permit requirements generally occurred with the BTEX components during upset conditions, and changes to the treatment process had reduced these occurrences.
- Accumulated effluent toxicity data from a number of acute and chronic tests using a variety of test species had demonstrated only low to moderate toxicity. The presence of a permitted mixing zone would further reduce toxicity outside of the regulated area.
- Long-term environmental monitoring results collected throughout the Port indicated that impacts to sediment chemistry and benthic communities were limited to the area near the effluent discharge point. In addition, monitoring of the intertidal organisms during the early years of the terminal operations when effluent concentrations were higher had not identified any impacts within these communities.

With these observations, we did not expect available data associated with the treated ballast water effluent to demonstrate an unacceptable chemical risk to ecological endpoints in the Port. However, other diverse sources may compound the potential stress caused to populations and communities by low-level, chronic hydrocarbon exposure associated with the BWTP, and the combined effects may be difficult to predict or understand (Lowell et al. 2000). Although this accumulation of stress through exposure to a complex set of stressors resulting from a variety of sources is the reality for most populations and communities, the traditional approach to EcoRA is only able to account for a limited fraction of this stress. We decided to take a nontraditional approach and to consider the gamut of environmental hazards possible in the Port. This decision added a regional perspective to the project resulting in a multiscaled assessment, including:

- A local scale that focused primarily on the BWTP effluent as a source and incorporated scientific data gathered for this purpose. The assessment completed at this scale followed the traditional EcoRA approach.
- A regional scale that focused on broad information available regarding the multiple sources and habitats in the Port and its surrounding watershed. Completing the assessment at this scale required modification to the EcoRA process as discussed in the following section.

LIMITATIONS OF TRADITIONAL RISK ASSESSMENTS AT THE REGIONAL SCALE

Typically, EcoRAs evaluate chemical concentration data with respect to single species toxicity data. In 1992, the EPA's EcoRA framework broadened this scope by discussing physical and biological stressors, as well as chemical stressors, and the importance of assessing multiple endpoints. More recently, guidance has emphasized larger scale or regional approaches, as evidenced by the merging of EcoRA with Watershed Assessments (Serveiss et al. 2000), and included cascading effects and cumulative impacts as necessary considerations when assessing whole ecosystems (USEPA 1997; 1998; 2003). Regardless of this trend, assessment goals and measurement endpoints are still mostly dependent on the dose–response relationship, and it is left to the risk assessor to try to integrate this simple relationship into the complex set of relationships that can exist within ecosystems.

To evaluate the range of information available for Port Valdez, we needed a larger, more inclusive data structure than was described in the 1992 EPA guidance available at the time. Once we had adjusted the scope of our information-gathering efforts, we then needed to modify the EcoRA process to address the following characteristics of the data set:

1. *Diverse Knowledge Base* — In order to broaden the information base and address ongoing community concern, we needed a method that could use traditional and anecdotal information, as well as scientific research.
2. *Systems Ecology* — The method needed to integrate information about stressors with the many interrelated components of the Port Valdez ecology and explore cumulative effects as a mechanism for potential decline in this system.
3. *Multiple Scales* — The method needed to integrate various exposure–effects relationships from a smaller-scale to a larger-scale evaluation.
4. *Long-term Management* — The method needed to act as an information management system that would assimilate new information and synthesize it with the old information. The information also needed to be in a form that could be reduced to easily understood conclusions about the state of the Port environment.

Modifications to the EcoRA approach resulted in the RRM. The model design is discussed in the next section, and the application to Port Valdez is described in the Methods and Results sections.

RELATIVE RISK MODEL DESIGN

The RRM design allowed us to extend the traditional EcoRA framework to provide a broad yet comprehensive screening assessment of impacts for all known sources in Port Valdez. The model design included the following steps:

- Categorization of eight source and habitat types in the region, and identification of potential ecological impacts expected from each source–habitat combination.
- Identification of three assessment endpoint categories based on public input, treating both scientific and anecdotal information equally.
- Delineation of 11 subareas based on the occurrence of habitat types, location of or transport potential from sources, and management concerns associated with assessment endpoints. Although the Port was the focus of the assessment, the subareas spanned the terrestrial, freshwater, and marine environment in recognition of the many interactions that occur between these areas.
- Conceptual site model development by defining the relationships of stressors and receptors to assessment endpoints within this structure.
- Development of criteria to rank the importance of the source and habitat categories between subareas. We based the ranking scheme on information that was readily available, could be consistently judged between subareas, and corroborated our understanding of likely risk factors from reviewing more detailed information about the Port.
- Calculation of relative risk by combining ranks for each subarea, weighted by the likelihood that the combination of a particular source and a particular habitat would result in an ecological impact.

The first step toward designing the model was to rescale the risk assessment components. Instead of focusing on specific stressors released into the environment and the receptors living in and using that environment, rescaling allowed us to focus on the sources releasing the stressors, and the habitats in which the receptors lived. At this scale, information was much easier to obtain and we were able to make assumptions about stressors when data were not available. For example, although hydrocarbons were a stressor of concern in the Port, the only chemical data available were associated with the BWTP and the city boat harbor. By rescaling the assessment, we were able to include the municipal wastewater treatment plant and contaminated runoff as potential sources of hydrocarbons.

Just as sources and habitats are more relevant at the regional scale than stressors and receptors, we also began to focus on the range of possible ecological impacts, rather than on individual receptor responses. Predicting the significance of ecological impacts is always the end goal of an EcoRA, but these predictions are made by extrapolating between levels of biological organization, and there is often little understanding of the implications of indirect effects (Preston 2002). At the regional scale, we concentrated on the physical prerequisites (e.g., spatial overlap of stressors and receptors, available transport pathways) for specific types of ecological impacts.

After identifying and categorizing the sources and habitats, we divided the study area into subareas based on groupings of these components. The subarea designations allowed us to use comparison (ranking) as a measuring technique. Ranking between subareas was an important tool in the RRM, because it normalized disparate data types and provided a semiquantitative measure based on concepts and qualifiers. For example, we ranked the subarea containing the BWTP higher than the subarea containing the municipal wastewater treatment plant because of the "larger effluent." This simple construction was easy to replicate for all sources and habitats.

Once we had completed these comparisons between subareas, we integrated the resulting information through a weighting process that screened out the less likely exposure pathways or impacted endpoints. This step is analogous to the risk characterization step of a traditional EcoRA where integration of information about exposure and effects forms the risk determination.

The RRM was beneficial in Port Valdez because it operated on qualitative and semiquantitative information and it provided a simultaneous analysis of the whole system. However, the regional-scale assessment is a relative measure of risk and does not specify the probability of an impact occurring. More detailed and quantitative determinations of risk were completed at the local scale (within subareas) to calibrate and confirm the regional model.

METHODS

The regional-scale assessment conformed to the three-phase approach of traditional risk assessments: *problem formulation, analysis,* and *risk characterization.* During the problem formulation, we gathered information from Port Valdez researchers, resource users, and residents. One of the essential elements of the problem formulation was a community meeting held in Valdez, Alaska to identify public concerns, values, and knowledge about the surrounding environment. We grouped the acquired information into categories relating to regional-scale risk components, which we then processed into an estimate of risk during the analysis phase, and interpreted during risk characterization to provide a comparative ecological risk perspective within the Port basin. We intended the results to inform stakeholders, not only of the chances of negative impacts associated with the oil industry, but also of the relative impacts from other anthropogenic uses and natural occurrences within the Port. This section describes the resources, decision points, and the means used to complete each phase of the assessment.

Problem Formulation

Background Investigation and Stakeholder Involvement

We initiated the investigation by asking three questions:

1. What are the physical and biological characteristics of the Port, including natural disturbances?
2. How do people interact with the environment?
3. What impacts are known to have occurred in the environment?

Baseline studies of the oceanographic and biological resources in Port Valdez provided information about seasonal fluctuations, circulation patterns, habitat types, and plant and animal populations. We examined various types of environmental discharge permits, determined if data regarding stressors were available, requested data when pertinent, and examined the literature to determine the range of stressors that could result from each source. The level of characterization varied for each source. Regulated and monitored sources, such as the NPDES-permitted facilities, were the most easily characterized, while characterization of other possible sources, such as contaminated runoff, consisted of generalized knowledge. Prior research efforts in the Port Valdez area and anecdotal information contributed to our understanding of the types of effects likely to occur in the Port.

We held three public meetings in the City of Valdez in October 1995 to aid in the formulation of assessment endpoints relevant to the Port. Following a brief introduction to the risk assessment process, the public was asked what concerned them about the Port Valdez environment. Responses were sorted into two categories: (1) stressors and sources of concern in the Port, and (2) populations or attributes of the Port that people wanted to protect. We also scheduled interviews in the community to supplement the public meetings and to ask specific questions that had arisen during the information-gathering phase. Participants included the city planning department, the Alaska Department of Environmental Conservation, and the U.S. Coast Guard (USCG), as well as local industry managers.

Assessment and Measurement Endpoints

Our discussions with risk managers, community interviews, and input from the public meetings resulted in selection of assessment endpoints. Fisheries, tourism, and the community's concern for the quality of its environment influenced the emphasis of the assessment endpoints. Each endpoint was also susceptible to one or more stressors possible in the Port Valdez environment. We defined the endpoint goals as assessing risk to the following areas:

1. Water and sediment quality in Port Valdez
2. Finfish and shellfish populations used by sport or commercial fishermen
3. Wildlife populations such as fishes, birds, and mammals that use the Port on either a year-round or seasonal basis

Assessment endpoints were carefully defined to reflect matters raised by resource managers and research scientists, as well as concerns voiced by the public (Wiegers et al. 1997). At times, these interests conflicted. For instance, a number of community members expressed concern that oil industry activities were affecting shellfish, and stated that they occasionally observed abnormal markings on crabs when harvesting shellfish. Scientific opinion suggested that crab populations dropped in the 1970s due to a growing sea otter population (Feder and Jewett 1988; Garshelis 1983). Another suggestion was that the yearly release of several hundred million hatchery fry increased feeding pressure on planktonic crab larvae. At this point in the project, we noted differing opinions, but this information did not influence the inclusion or exclusion of an endpoint. We also discussed possible measurement endpoints that would aid in the evaluation of the assessment endpoints, an important consideration during data review and hypothesis testing.

Results of the Problem Formulation: Conceptual Model

Information gathered during the problem formulation phase provided the foundation for constructing the conceptual model. Initially, we focused on describing the standard components of a risk assessment: stressors, receptors, and the direct and indirect effects that could result from the interaction of the first two components. This information was regrouped into categories relevant to the regional-scale risk assessment components of sources, habitats, and ecological impacts. Source and habitat categories describe the anthropogenic and ecological components of the Port (Table 4.1). Impact categories described the chosen assessment endpoints. We then divided the Port into 11 separate subareas. The locations and boundaries of each subarea are described in Table 4.1 and illustrated in Figure 4.1.

Once the regional-scale categories were established, we explored exposure and effect characteristics for each combination of components by developing working tables for each subarea. The tables summarized information that would affect exposure, such as temporal or spatial distribution of typical stressors and receptors, and that would affect receptor responses, such as life stages and community interactions. Based on the information organized in the tables, we were able to conceptualize generalized risk scenarios for each subarea. This approach ensured that we were informed about and had considered the interaction of individual stressors and receptors before making professional judgments on the regional scale. The risk scenarios also provided a conceptual structure from which to develop hypotheses for future quantitative assessments.

Analysis

The table-based structure of the conceptual model simulated general aspects of the Port and provided a single framework within which to formulate risk scenarios. The analysis phase of the assessment included two approaches: comparative analysis of risks at a regional scale and quantitative analyses of site-specific risk using traditional risk assessment techniques. We also addressed uncertainty and sensitivity during the relative risk analysis.

Relative Risk Model

The RRM compared the 11 subareas of interest in order to determine where the presence of multiple sources and sensitive habitats is more likely to affect assessment endpoints. The model design for Port Valdez makes the following assumptions:

1. The greater the size or frequency of a source in a subarea, the greater the potential for exposure to stressors.
2. The type and density of receptors present is related to the available habitat.
3. The sensitivity of receptors to stressors varies in different habitats; the severity of effects between different subareas of the Port depends on relative exposures and the characteristics of the receptors present.

As described in Chapter 2, the resulting model is a system for ranking risk components and filtering each possible combination to arrive at a reasoned and

Table 4.1   Subareas, Sources, and Habitats Defined for the Port Valdez Ranking Risk Assessment

Subareas (Risk Regions)

Shoup Bay
Shoup Bay, including the bay entrance, the entrance spit, and a portion of the shoreline to the east of the bay

Mineral and Gold Creeks
Shoreline area and the shallow shelf of the Mineral Creek embayment, including Gold Creek

City of Valdez
The city and the shoreline and shallow shelf areas from just east of Mineral Creek to the eastern end of the Small Boat Harbor

Duck Flats (or Mineral Island Flats) and Old Valdez
The Duck Flats, including the islands and shallow shelf south of the flats, and the shoreline area including the Richardson Highway extending east to the Valdez Glacier Stream

Robe and Lowe Rivers
Shoreline, river deltas, and shallow subtidal areas of the Valdez Glacier Stream, Robe River and Lowe River, including the Petro Star Refinery

Dayville Flats and Solomon Gulch
Shoreline along Dayville Road and shallow subtidal areas from the southern edge of the Lowe River to just east of Allison Point, including the Solomon Gulch Hatchery

Valdez Marine Terminal
Shoreline and shallow subtidal areas from Allison Point to just west of Saw Island, including the Valdez Marine Terminal

Sawmill to Seven-Mile Creeks
Shoreline and shallow subtidal areas from west of Saw Island to a point east of Anderson Bay, including Sawmill Creek, Five-Mile Beach, and Seven-Mile Beach

Anderson Bay
Shoreline and shallow subtidal areas from just east of Anderson Bay to the west of Entrance Island

Western Port
The western, flat-bottomed basin from the Valdez Narrows to a middle boundary between the Mineral Creek embayment to the eastern edge of the Valdez Marine Terminal

Eastern Port
The eastern, upward-sloping basin from the middle boundary to the edge of the shallow offshore area of the eastern shoreline

Sources

Treated Discharges
Effluents from point sources (released from a pipe) that are treated to reduce chemical and physical contaminants before release

Contaminated Runoff
Runoff from land that has been contaminated through air pollution, groundwater contamination, spills on land, pesticide and other chemical applications, or another process

Accidental Spills
Spills of oil, lubricants, solvents, antifreeze, fluids, or other chemicals on the water

Fish and Seafood Processing Wastes
Wastes composed of solid or settling organic matter, including seafood processing, sport fish wastes, and food or fecal matter resulting from aquatic culturing

**Table 4.1   Subareas, Sources, and Habitats Defined for the Port Valdez Ranking
            Risk Assessment (continued)**

Vessel Traffic
Small or large vessels that may cause injury through contact or propeller wash, disturbance
from noise or movement, release of fuels and other chemicals from normal operation, release
of sewage wastes, or release of ballast water

Construction and Development
Activities such as land clearing, building, and road and dock construction that directly alter
habitat, release debris or sediment, or change physical conditions such as water flow

Hatchery Fish
Salmon returning to the hatchery that stray into other spawning streams, and hatchery fry
migrating out of the port

Shoreline Activity
Recreational or residential activity resulting in disturbance or injury

Habitats

Saltmarsh
Shoreline areas characterized by marsh grasses and sedges

Mudflats
Shoreline areas with an extensive tidal flat consisting of mostly silt and clay sediments

Spits and Low-Profile Beaches
Flat shoreline areas or spits extending out from the shoreline that consist of broken rock, cobble
beaches, or coarse sediment and gravel

Rocky Shoreline
Sloped to steep shorelines consisting of large rocks, boulders, or seacliffs

Shallow Subtidal
Water column and benthic areas less than 50 m deep with either sediment or rocky bottoms

Deep Benthic
Underwater areas greater than 50 m deep consisting of mostly a sediment bottom

Open Water
Water column or pelagic zone in deep water areas where influences from land are lessened

Stream Mouths
Intertidal mud, sandy gravel, and gravel entrances to streams and rivers and upstream areas
influenced by tidal flows

repeatable estimate of relative risk. Application of this system to Port Valdez involved
the following.

Ranking

Sources and habitats in each subarea were ranked to indicate a relative probability
(low, medium, or high) that assessment endpoints could be significantly impacted.
Criteria were based on the size and frequency of the source and the amount and use
of available habitat. Uncertainty associated with each criterion was also described.
The ranking criteria for each variable are presented in Table 4.2. The resultant
ranking values are provided in Table 4.3.

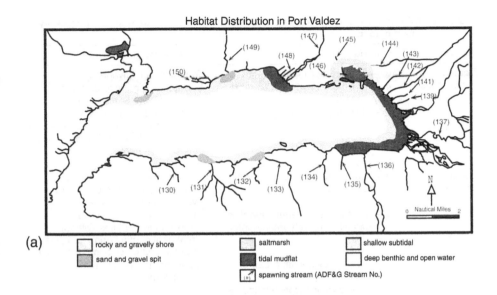

(a)

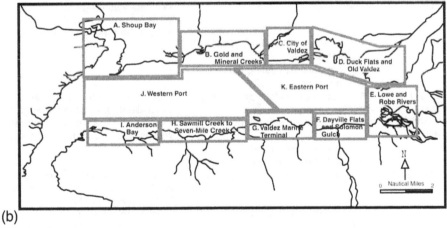

(b)

Figure 4.1   Habitat types and subarea (risk region) delineations chosen for the analysis of Port Valdez. Detailed descriptions are given in Table 4.1.

Filter Design

Exposure and effect filters were designed to characterize the relationship between risk components (sources, habitats, and impacts to assessment endpoints) and consisted of a table of weighting factors for the component combinations of interest. A single-exposure filter was designed for the source and habitat combinations in Port Valdez. The design of the effect filter was similar, but a separate filter was made for each assessment endpoint. The exposure filters and the effects filters are provided in Table 4.4.

Table 4.2   Criteria for Ranking Sources and Habitats: Factors Leading to Uncertainty Are Included

Source	Criteria	Uncertainty in the Criteria
Treated discharges	6 — flow greater than 10 mgd	Treatment effectiveness
	4 — flow between 5 and 10 mgd	Undetected sporadic discharge of contaminants at high levels
	2 — flow less than 5 mgd	Continuous discharge of contaminants below detection levels, especially for contaminants that can accumulate in the environment
	0 — no flow	
Contaminated runoff	6 — large industrial, commercial, or dense residential areas	Some sites have stormwater containment and treatment (e.g., Valdez Marine Terminal)
	4 — light industrial areas, landfills, or subdivisions with septic tanks	Contamination in stormwater from storm drains or sites without treatment or monitoring (e.g., the city, most industrial or commercial sites)
	2 — sparse residential areas or possible mining	Contaminated runoff from active and inactive mines
	0 — no known or suspected sources of contamination	
Accidental spills	6 — loading or unloading facilities for fuels or oil	Spills at sites that are highly monitored (e.g., the Valdez Marine Terminal and other fuel transfer docks) are more likely to be reported and cleaned up
	4 — other docks or commercial boating activity	
	2 — recreational boating activity	
	0 — no sources of spills	
Fish and seafood processing wastes	6 — seasonal seafood processing waste streams	Dispersal on the bottom depends on water depth and current strengths
	4 — seasonal use of net pens	Some organic solids may contain other wastes (e.g., cleaners, antibiotics)
	2 — sporadic fish wastes	
	0 — no known or suspected sources	
Vessel traffic	6 — year-round daily traffic present	Commercial shipping, especially for crude oil, is frequent, although long-term trends may change
	4 — year-round monthly traffic present	Recreational, charter and tour services, and fishing traffic are seasonal and may be sporadic
	2 — seasonal traffic	
	0 — little boat traffic expected	
Construction and development	6 — large-scale development expected	Construction activities are mostly seasonal and short term, although a specific project may last over years
	4 — frequent construction or small-scale development expected	Areas where future development projects are planned have high uncertainty
	2 — developed	
	0 — no current or expected development	
Hatchery fish	6 — near hatchery	The number of hatchery fish that stray into other streams is not known

Table 4.2   Criteria for Ranking Sources and Habitats: Factors Leading to Uncertainty Are Included (continued)

Source	Criteria	Uncertainty in the Criteria
	4 — expected adult and fry migration route	The criteria assume straying is more likely on the southern shore near the hatchery
	2 — possible locations of adult and fry	
	0 — no hatchery fish expected	
Shoreline activity	6 — daily activity, year round	Exposure depends on type of activity, proximity to receptors, and sensitivity of the receptors
	4 — recreational, road access	Some receptors occur or are more sensitive on a seasonal basis (e.g., migratory birds, spawning salmon)
	2 — recreational, no road access	
	0 — little shoreline activity expected	

Habitat	Criteria	Uncertainty in the Criteria
Mudflats	6 — extensive mudflats	Population density and community types vary depending on sediment grain size, nutrient and organic carbon levels, sedimentation, and salinity
	4 — moderate or extensive mudflats with low population densities	
	2 — limited mudflat areas	
	0 — no mudflats	
Saltmarsh	6 — extensive saltmarsh	High productivity of saltmarshes and infrequent occurrence of this habitat type in Prince William Sound may increase its regional importance
	4 — moderate area of saltmarsh	Disturbance would affect some populations more than others (e.g., high-use habitat for migratory birds)
	2 — limited saltmarsh areas	
	0 — no saltmarsh	

Habitat	Criteria	Uncertainty in the Criteria
Spits and low-profile beaches	6 — spits, spit-like formations, or extensive low-profile beaches	Generally low productivity may limit the importance of this habitat type
	4 — some low-profile beaches	Importance of these areas may depend on their proximity to other habitats
	2 — limited areas with low-profile beaches	
	0 — no spits or low-profile beaches	
Rocky shoreline	6 — extensive rocky shoreline	Population density and community types vary depending on the availability of nutrients and organic carbon, sedimentation, salinity, and wave action
	4 — some rocky shoreline	
	2 — limited rocky shoreline	
	0 — no rocky shoreline areas	

Table 4.2 Criteria for Ranking Sources and Habitats: Factors Leading to Uncertainty Are Included (continued)

Shallow subtidal (< 50 m deep)	6 — extensive shallow subtidal shelf	Limited or narrow areas of shallow subtidal in the Port
	4 — moderate shallow subtidal area	This habitat group does not differentiate between hard- and soft-bottomed subtidal areas, which will affect the biological activity in the habitat
	2 — narrow shallow subtidal area	
	0 — no shallow subtidal areas	
Deep benthic (> 50 m deep)	6 — extensive deep subtidal areas	Population density and community types are affected by the amount of settling sediment and occasional seismic slumping
	4 — moderate deep subtidal areas	Sediment grain size, which varies slightly in the eastern and western Port, also influences animal assemblages
	2 — limited deep subtidal areas	
	0 — no deep subtidal areas	
Open water	6 — large areas with deep water column	Flushing in the Port is tied to seasonal events, variability in the tides and currents, and stratification of the water column
	4 — moderate areas with deep water column	Nutrient cycling in the Port is related to stratification of the water column and to yearly variation in phytoplankton and zooplankton communities
	2 — small areas with deep water column	
	0 — no deep water	
Stream mouths	6 — large river or creek systems with many freshwater tributaries	Steep terrestrial slopes of Port Valdez limit stream habitat areas
	4 — streams with few tributaries, moderate flows	Stream mouths are exposed to large variations in salinity and turbidity, substrate found at stream mouths is coarser than most sediments in the Port
	0 — no streams	

Integrating Ranks and Filters

Ranks and weighting factors were combined through multiplication. The results formed a matrix of risk scores related to the relative exposure or effects associated with a source and habitat in each subarea. Summing by subarea results in the relative estimate for each subarea.

Uncertainty Analysis

In this study, we addressed uncertainty (1) in the conceptual model, (2) in the calculation of relative risk, and (3) in the accuracy of relative risk estimates in Port Valdez. Uncertainty associated with the structure of the conceptual model was mostly qualitative. The calculation of relative risk had a quantifiable level of uncertainty.

Table 4.3 Input to Relative Risk Model: Ranking for Source and Habitat by Subareas

	Source Ranks							
Subarea	Treated Discharge	Contaminated Runoff	Accidental Spills	Fish Waste	Vessel Traffic	Construction Development	Hatchery Fish	Shoreline Activity
Shoup Bay	0	2	2	0	2	0	0	2
Mineral and Gold Creeks	0	2	2	0	2	2	0	4
City of Valdez	0	6	6	6	6	4	0	6
Duck Flats and Old Valdez	4	4	4	0	4	4	0	6
Lowe and Robe Rivers	0	4	2	0	2	2	2	2
Dayville and Solomon Gulch	0	2	4	4	4	4	6	4
Valdez Marine Terminal	6	4	6	2	6	4	4	6
Sawmill to Seven-Mile Creeks	0	0	2	0	2	0	4	0
Anderson Bay	0	0	2	0	2	6	4	2
Western Port	0	0	2	2	6	0	0	0
Eastern Port	6	0	2	2	4	0	0	0

	Habitat Ranks							
Subarea	Mudflat	Saltmarsh	Spits and Beaches	Rocky Shore	Shallow Subtidal	Deep Benthic	Open Water	Stream Mouth
Shoup Bay	2	0	6	6	4	4	4	2
Mineral and Gold Creeks	4	0	2	4	6	0	0	6
City of Valdez	0	0	4	2	4	0	0	0
Duck Flats and Old Valdez	6	6	0	4	6	0	0	6
Lowe and Robe Rivers	6	0	0	0	2	0	0	6
Dayville and Solomon Gulch	4	0	2	4	2	0	0	4
Valdez Marine Terminal	2	0	2	2	2	0	0	2
Sawmill to Seven-Mile Creeks	2	0	6	2	2	0	0	2
Anderson Bay	2	0	2	6	2	0	0	2
Western Port	0	0	0	0	0	6	6	0
Eastern Port	0	0	0	0	0	6	6	0

Table 4.4 Inputs to the Relative Risk Model: Filters for Exposure from Each Source to Each Habitat and for the Effects for Each Endpoint under Evaluation

Habitats	Treated Discharge	Contaminated Runoff	Accidental Spills	Fish Waste	Vessel Traffic	Construction Development	Hatchery Fish	Shoreline Activity
Exposure Filter								
Saltmarsh	0	1	1	0	0	1	0	1
Mudflat	0	1	1	0	0	1	0	1
Spits and Beaches	0	1	1	0	0	1	0	1
Rocky Shoreline	0	0	1	0	0	0	0	1
Shallow Subtidal	1	1	1	1	1	1	0	0
Deep Benthic	1	0	0	1	1	1	0	0
Open Water	1	1	1	0	1	0	1	0
Stream Mouth	0	1	1	0	0	1	1	0
Effects Filter: Water Quality								
Saltmarsh	0	1	1	0	0	1	0	0
Mudflat	0	1	1	0	0	1	0	0
Spits and Beaches	0	1	1	0	0	1	0	0
Rocky Shoreline	0	0	1	0	0	0	0	0
Shallow Subtidal	1	1	1	1	1	1	0	0
Deep Benthic	1	0	0	1	1	1	0	0
Open Water	1	1	1	0	1	0	0	0
Stream Mouths	0	1	1	0	0	1	1	0
Effects Filter: Sediment Quality								
Saltmarsh	0	1	1	1	0	1	0	1
Mudflat	0	1	1	1	0	1	0	1
Spits and Beaches	0	1	1	1	0	1	0	1
Rocky Shoreline	0	0	1	0	0	0	0	1
Shallow Subtidal	1	1	1	1	1	1	0	0
Deep Benthic	1	0	0	1	1	1	0	0

Open Water	1	1	1	0	1	0	1	0
Stream Mouths	0	1	1	0	0	1	1	0
Effects Filter: Hatchery Salmon Culture and Migration								
Saltmarsh	0	1	1	0	0	1	0	0
Mudflat	0	1	1	0	0	1	0	0
Spits and Beaches	0	1	1	0	0	1	0	0
Rocky Shoreline	0	0	1	0	0	0	0	0
Shallow Subtidal	1	1	1	1	1	0	0	0
Deep Benthic	0	0	0	0	0	0	0	0
Open Water	1	1	1	0	1	0	1	0
Stream Mouths	0	1	1	0	0	1	1	0
Effects Filter: Bottom Fishes and Shellfishes								
Saltmarsh	0	0	0	0	0	0	0	0
Mudflat	0	0	0	0	0	0	0	0
Spits and Beaches	0	0	0	0	0	0	0	0
Rocky Shoreline	0	0	1	0	0	0	0	0
Shallow Subtidal	1	1	1	1	1	0	0	0
Deep Benthic	1	0	0	1	1	1	0	0
Open Water	0	0	0	0	0	0	0	0
Stream Mouths	0	0	0	0	0	0	0	0
Effects Filter: Wild Anadromous Fishes								
Saltmarsh	0	1	1	0	0	1	0	0
Mudflat	0	1	1	0	0	1	0	0
Spits and Beaches	0	1	1	0	0	1	0	0
Rocky Shoreline	0	0	1	0	0	0	0	0
Shallow Subtidal	1	1	1	1	1	0	0	0
Deep Benthic	0	0	0	0	0	0	0	0
Open Water	1	1	1	0	1	0	1	0
Stream Mouths	0	1	1	0	0	1	1	0

Table 4.4　Inputs to the Relative Risk Model: Filters for Exposure from Each Source to Each Habitat and for the Effects for Each Endpoint under Evaluation (continued)

Effects Filter: Bird Reproduction

Habitats	Sources							
	Treated Discharge	Contaminated Runoff	Accidental Spills	Fish Waste	Vessel Traffic	Construction Development	Hatchery Fish	Shoreline Activity
Saltmarsh	0	1	1	0	0	1	0	1
Mudflat	0	1	1	0	0	1	0	1
Spits and Beaches	0	1	1	0	0	1	0	1
Rocky Shore	0	0	1	0	0	0	0	1
Shallow Subtidal	1	1	1	0	1	0	0	0
Deep Benthos	0	0	0	0	0	0	0	0
Open Water	1	1	1	0	1	0	0	0
Stream Mouths	0	1	1	0	0	1	0	0

Effects Filter: Food Availability for Wild Fishes, Birds, and Mammals

Habitats	Treated Discharge	Contaminated Runoff	Accidental Spills	Fish Waste	Vessel Traffic	Construction Development	Hatchery Fish	Shoreline Activity
Saltmarsh	0	1	1	0	0	1	0	1
Mudflat	0	1	1	0	0	1	0	1
Spits and Beaches	0	1	1	0	0	1	0	1
Rocky Shoreline	0	0	1	0	0	0	0	1
Shallow Subtidal	1	1	1	1	1	0	0	0
Deep Benthic	1	0	0	1	1	0	1	0
Open Water	1	1	1	0	1	1	1	0
Stream Mouths	0	1	1	0	0	1	1	0

We designed a sensitivity analysis to ascertain the variance of the results associated with the mathematical model and the modeling input. Accuracy of the relative risk results was explored through comparison of the confirmatory analyses used to quantify or describe specific risks in the Port.

Sensitivity Analysis

The sensitivity analysis included two phases. Initially, the factors driving the model were investigated by running the model with limited components. During the second phase, we incorporated randomly chosen input and examined the results for each subarea. We ran an additional test to determine the sensitivity of the model when uncertainty in the ranks was considered. Instead of using randomly chosen ranks for the input values, we allowed the model to choose from within a range of ranks representing our uncertainty in the ranked values used for Port Valdez. The ranges below were our subjective estimates of the probability and associated uncertainty of impacts occurring, which we applied to each source–habitat combination:

0	none (or very unlikely)
0 to 2	unlikely
0 to 4	unlikely but somewhat uncertain
0 to 6	possible but very uncertain
2 to 6	possible and somewhat uncertain
4 to 6	likely

We ran 20 trials with the randomly selected input. The results from these analyses were plotted to demonstrate the possible variation in the results of the RRM when uncertainty was included in the ranking process. The effect filters were not examined in the sensitivity analysis as they were expected to have a similar influence on the model results as the exposure filters.

Confirmatory Analysis

Available chemical data from Port Valdez provided an opportunity to test the results of the RRM with more traditional analyses of risk from specific stressors. Two approaches were used for the confirmatory analyses: (1) comparison of chemical concentrations in effluent, sediment, and tissue samples to benchmark values; and (2) modeling of chemical concentrations in sediment samples to determine toxicity to marine amphipods. Each approach focused on chemical exposure and effects; available data were not sufficient to assess physical or biological stressors in a similar manner.

Benchmark Values

This analysis compared PAH and metal concentrations from Port Valdez samples to threshold levels derived in the literature. The Port data were compiled from samples collected in conjunction with the BWTP permit (Alaska Pipeline Service Company), the Alyeska Environmental Monitoring Program (Feder and Shaw 1993a;

1993b; 1994a; 1995; 1996), the Long-Term Monitoring Program (LTEMP) (Kinnetics Laboratories 1995; 1996), and the U.S. Army Corps of Engineers sampling in the small boat harbor (U.S. Army Corps of Engineers, 1995), and a sea otter disturbance study (Anthony 1995). Benchmark values were derived from the U.S. EPA (USEPA 1996) program for developing ecotox thresholds (ETs), freshwater benchmarks developed by Suter (1996), sediment effect ranges set by the National Oceanic and Atmospheric Administration (NOAA) and developed by Long and Morgan (1990), and wildlife threshold levels developed by Opresko et al. (1995). The purpose of each study was to synthesize effect-based data into useful criteria for determining the levels at which adverse effects occur. We compared the benchmark values to PAH and metal concentrations in sediments, effluent, and mussel tissue from various locations in the Port and tallied the number of times each sample concentration exceeded benchmark values.

Modeling PAH Toxicity in Sediments

The concentrations of selected PAHs in the sediments of Port Valdez have been collected in a number of monitoring studies and occasional sampling events. Sampling data included in this analysis are the same as those used in the benchmark analysis above: small boat harbor (U.S. Army Corps of Engineers 1995), offshore of the Valdez Marine Terminal and Gold Creek (Feder and Shaw 1993a; 1993b; 1994a; 1995; 1996; Kinnetics Laboratories 1995; 1996), near Solomon Gulch Hatchery (Shaw 1996), and other deep water areas of the Port (Feder and Shaw 1993b; 1994a; 1995; 1996).

These measured values provided input for the ΣPAH model developed by Swartz et al. (1995). The model combines the following five well-known models that can be applied to hydrocarbons in sediment.

1. Equilibrium Partitioning Model: describes the partitioning of PAH in the sediment interstitial water based on the total organic carbon content of the sediments.
2. QSAR Model: determines the acute toxicity of individual PAHs to amphipods in a 10-day test.
3. Toxic Unit Model: describes the toxicity of PAHs in interstitial water.
4. Additivity Model: determines the total toxicity from 13 selected PAHs.
5. Concentration–Response Model: describes the mortality response of amphipods to spiked field sediments.

The ΣPAH model predicts the probability of no toxicity (defined as < 13% mortality), uncertain toxicity (defined as 13 to 24% mortality), and toxicity (defined as > 24% toxicity).

RESULTS

Relative Risk in Port Valdez

Systematic application of the conceptual model to the habitats and risk sources in each of the subareas led to a ranking of relative risk within the Port environment.

The risk scores are unitless numbers that judge the relative severity of environmental risk based on an informed decision-making process. The relative risk scores for Port Valdez are presented in Tables 4.5 and 4.6, and summarized in Figure 4.2. The total relative risk for each subarea in Port Valdez was calculated by summing across the rows in either of the matrices in Table 4.5 or Table 4.6.

The scores ranged from 40 (Sawmill to Seven-Mile Creeks) to 448 (Duck Flats and Old Valdez). We considered subareas with scores less than 150 to have low relative risk. Subareas in this group included Shoup Bay, Sawmill to Seven-Mile Creeks, Anderson Bay, and the Western Port. Subareas with scores between 150 and 300 were considered to have moderate relative risk. These included Mineral and Gold Creeks, City of Valdez, Robe and Lowe Rivers, Dayville Flats and Solomon Gulch, and the Valdez Marine Terminal. Only one subarea, Duck Flats and Old Valdez, had a high risk score greater than 300. Because of the uncertainty associated with the ranking process, comparisons of relative risk more detailed than these low, moderate, and high groupings are probably not meaningful.

Our analysis suggested that the pelagic environment and western shoreline, areas affected by less development, are at low relative risk. Most of the eastern shoreline is at moderate relative risk. This includes subareas from the City of Valdez to the Valdez Marine Terminal where a variety of development has occurred. The one subarea of high relative risk, Duck Flats and Old Valdez, is located in the developed eastern area. The greater risk predicted here by the model is related to the diversity and quality of habitats in this area. Note that "high relative risk" may or may not imply high risk in an absolute sense. Instead, this suggests that a greater degree of environmental stress is possible, and that there is a higher probability that significant ecological impacts will occur than in other areas of the Port.

The contribution of the eight-stressor sources to relative risk in the entire Port Valdez region can be determined by summing down the column in Table 4.5. Applying the same criteria defined above (low relative risk < 150; moderate relative risk, 150 to 300; and high relative risk > 300), treated discharges, fish and seafood wastes, and the presence of hatchery fish rank as low relative risk; vessel traffic and construction and development activities as moderate relative risk; and contaminated runoff, accidental spills, and shoreline activity as high relative risk. This distribution of relative risk between sources is reasonable when characteristics of the stressors associated with these sources are considered. Runoff, spills, and shoreline activity behave similarly to nonpoint discharges, and the effects are likely to be widely distributed throughout the Port. Treated discharges, and fish and seafood wastes behave more like point sources and may be discharged into fewer subareas and possibly less sensitive habitats.

The contribution of the eight habitat categories to relative risk in Port Valdez as a whole can be determined by summing down the columns in the second matrix of Table 4.6. Using the same criteria defined above, saltmarsh and deep benthic habitats rank as low relative risk; spits and low-profile beaches, the rocky shoreline, and open water habitats as moderate relative risk; and mudflats, shallow subtidal, and stream mouth habitats as high relative risk. Relative risk to habitats in Port Valdez as a whole is strongly influenced by the abundance of habitats across subareas. For instance, saltmarsh occurs in only one subarea, Duck Flats and Old Valdez. Although

Table 4.5 Ranked Relative Risk Output of Model by Source and Subarea. (The far right column is the sum for each subarea; the bottom row is the sum for each source type)

Subarea	Sources								Total Relative Risk
	Treated Discharge	Contaminated Runoff	Accidental Spills	Fish Waste	Vessel Traffic	Construction Development	Hatchery Fish	Shoreline Activity	
Shoup Bay	0	36	48	0	24	0	0	28	136
Mineral and Gold Creeks	0	36	44	0	12	24	0	40	156
City of Valdez	0	48	60	24	24	16	0	36	208
Duck Flats and Old Valdez	24	96	112	0	24	72	0	96	424
Lowe and Robe Rivers	0	56	28	0	4	48	12	12	160
Dayville and Solomon Gulch	0	24	48	8	8	20	24	24	156
Valdez Marine Terminal	12	32	72	0	12	24	8	48	208
Sawmill to Seven-Mile Creeks	0	0	28	0	4	0	8	0	40
Anderson Bay	0	0	28	0	4	36	8	20	96
Western Port	0	0	24	12	72	0	0	0	108
Eastern Port	72	0	24	12	48	0	0	0	156
Total Relative Risk	108	328	516	56	239	240	60	304	

Table 4.6 Ranked Relative Risk Output of Model by Habitat and Subarea. (The far right column is the sum for each subarea [risk region]; the bottom row is the sum for each habitat type)

Habitats Subarea	Mudflat	Saltmarsh	Spits and Beaches	Rocky Shoreline	Shallow Subtidal	Deep Benthic	Open Water	Stream Mouth	Total Relative Risk
Shoup Bay	12	0	36	24	24	8	24	8	136
Mineral and Gold Creeks	40	0	20	24	36	0	0	36	156
City of Valdez	0	0	88	24	96	0	0	0	208
Duck Flats and Old Valdez	108	108	0	40	96	0	0	72	424
Lowe and Robe Rivers	72	0	0	0	16	0	0	72	160
Dayville and Solomon Gulch	48	0	24	0	28	0	0	56	156
Valdez Marine Terminal	40	0	40	48	48	0	0	36	208
Sawmill to Seven-Mile Creeks	4	0	12	4	8	0	0	12	40
Anderson Bay	20	0	20	24	8	0	0	24	96
Western Port	0	0	0	0	0	48	60	0	108
Eastern Port	0	0	0	0	0	72	84	0	156
Total Relative Risk	344	108	240	188	360	128	168	316	

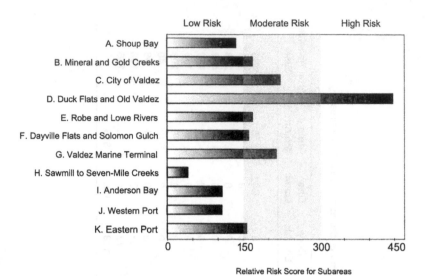

Figure 4.2   Total relative risk scores obtained for each subarea as categorized as high, medium, and low risk.

saltmarsh receives the highest possible ranking *in that subarea*, that alone still leads to a low relative risk to Port Valdez *as a whole*. The reverse situation occurs for open water habitat. The risk to open water in any individual subarea is never more than half the maximum possible, but open water occurs in every subarea. The result is that open water habitats have high relative risk for Port Valdez as a whole.

These results were based on the source–habitat combinations with the exposure filter only applied. The effect filters further refined the scores and developed more specific results regarding each assessment endpoint. Figure 4.3 shows the distribution of relative risk (a) across the Port with the exposure filter alone, and (b) with the exposure filter and the water quality effect filter. The results were similar, except that the relative risk in two subareas (Gold and Mineral Creeks and Dayville Flats and Solomon Gulch) changed from moderate to low with the water quality filter applied. This difference reflected the removal of shoreline activity as a source of concern for water quality issues.

Uncertainty

The features of the relative risk assessment gave rise to five general sources of uncertainty:

1. *Missing Information*: Information gaps occur where sources or stressors in the Port were not identified or important aspects of the ecology were not developed.
2. *Ambiguities in the Available Information*: Ambiguity exists in the anecdotal, regulatory, and scientific data collected regarding the purposes of this study.
3. *Error in the Conceptual Model*: The conceptual model defines the components and the links between these components that contribute to risk in the Port Valdez system. Undefined links or links interpreted incorrectly will cause errors in accuracy or precision of the relative risk descriptions.

(a) Exposure to Stressors

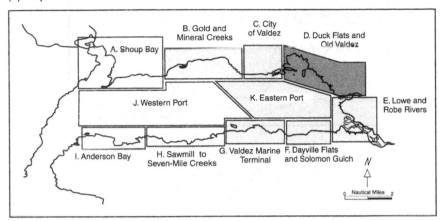

(b) Water Quality

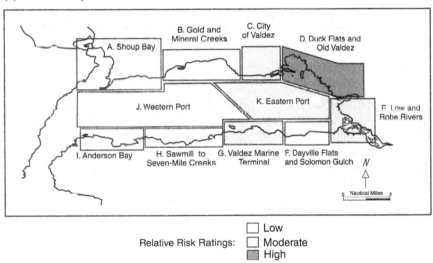

Relative Risk Ratings: ☐ Low
☐ Moderate
■ High

Figure 4.3   Relative risks associated with (a) exposure and (b) impacts to water quality.

4. *Error in the Estimate of Relative Risk*: Misconceptions in the decision-making process or inaccuracies in the numerical processing could result in erroneous results. This error is partially evaluated through the sensitivity analysis.

5. *Variability in the Environment:* The combination of nonlinear and stochastic properties of nature creates variability in plant and animal populations and causes variable responses to stressors. This form of uncertainty can be described, but not reduced.

We assume that the estimates of ecological risk to Port Valdez derived from our conceptual model contain substantial uncertainty. This uncertainty is reflected in our categorizing relative risk in the broad terms of low, moderate, and high.

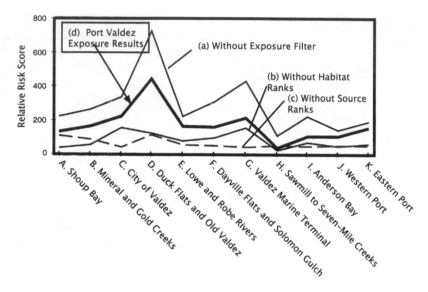

Figure 4.4   Model results when driven by (a) source and habitat ranks, (b) source ranks and
the exposure filter, (c) habitat ranks and the exposure filter, and (d) all of these
components.

Sensitivity

The sensitivity of the RRM is related to its ability to identify the difference
between high- and low-risk areas. To analyze the sensitivity of this model, we
incorporated randomly chosen input and examined the results for each subarea. The
sensitivity analysis is based on the premise that when input is randomly chosen, the
model results will not discriminate between different subareas of the Port. Input that
is risk related, instead of random, will drive the model to detect the high-risk areas.

The first phase of the sensitivity analysis investigated the components (source
ranks, habitat ranks, and filter values) driving the model output. In Figure 4.4, the
relative risk results are compared to results when one of the components is removed
from the analysis. When the exposure filter was removed from the model, the results
varied primarily in magnitude (Figure 4.4a). The exposure filter appeared to have
little effect on the comparative results: the Duck Flats and Old Valdez subarea still
received the highest relative risk score. However, the ranks affected the model in a
different manner than the filter. When habitat ranks were excluded from the model
input, the sources drove the analysis, resulting in the City of Valdez and Valdez
Marine Terminal subareas receiving the highest scores (Figure 4.4b). When source
ranks were excluded, the habitats drove the analysis, and Shoup Bay and the Duck
Flats and Old Valdez subareas received the highest relative risk scores (Figure 4.4c).
When all of the input was included, the model combined characteristics of the source,
habitat, and exposure components to produce the final result (Figure 4.4d).

Uncertainty in the ranking process depends on the accuracy of the ranks chosen
to represent risk from source to habitat. To explore uncertainty in these choices, we
established a range of values around each rank originally used in the relative risk

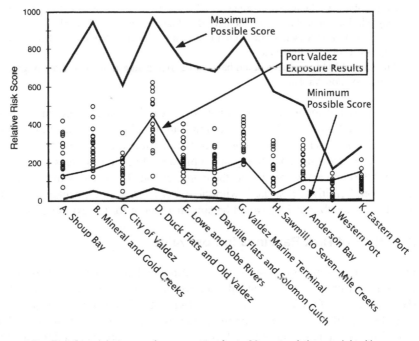

Figure 4.5   Relative risk scores for exposure from 20 runs of the model with source and habitat ranks randomly drawn from within a specified range of values representing uncertainty in the ranks. The upper and lower values are the maximum and minimum results possible in this uncertainty analysis. The middle line is the result from the relative risk analysis.

analysis. The range was limited to numbers used in the ranking process: 0, 2, 4, and 6. High values represented the highest rank expected for the source or habitat in the subarea. Low values represented the lowest rank expected. These values were chosen conservatively, so even a slight uncertainty was represented within the range. Single values were only used when little doubt existed that the source or habitat was not present in the subarea. A zero was assigned in these cases.

To test sensitivity and uncertainty, we ran the model 20 times with the following sets of input: (1) all rank and exposure filter values chosen randomly; and (2) ranks chosen from within the specified uncertainty ranges and the exposure filter set at the original values. Allowing only random input into the model produced the least sensitive result. In this case, the subarea most frequently identified as having the highest relative risk was Lowe and Robe Rivers. The model detected this result five times in 20 runs, or 25% of the time. When the ranked input was limited by the range of possible values and the exposure filter was not random the model detected the same high-risk area (Duck Flats and Old Valdez) more than 55% of the time. This result, with uncertainty incorporated, agreed with the result from the initial relative risk analysis where uncertainty was not considered.

Figure 4.5 compares the original risk analysis results to the 20 sets of results obtained through the uncertainty analysis, as well as the highest and lowest possible results of the uncertainty analysis. The two most notable differences were:

1. Most subareas (Shoup Bay, Mineral and Gold Creeks, Lowe and Robe Rivers, Dayville Flats and Solomon Gulch, Valdez Marine Terminal, Sawmill to Seven-Mile Creek, and Anderson Bay) could be at a higher risk than established in the original results for Port Valdez.
2. The City of Valdez subarea could be at lower risk than established in the original results for Port Valdez.

In summary, the model is most sensitive to a reduction in uncertainty in the source and habitat ranks. Reducing uncertainty in the filter would have little effect on the comparative results of the model, but would affect the magnitude of the final scores. However, reducing uncertainty in the process of ranking sources and habitats would affect the comparative results and the identification of subareas with high relative risk in the Port. Sensitivity associated with the effect filters was not explored here, but we assume that the observations would be similar to those for the exposure filter.

Confirmation of Risk Rankings in Port Valdez

The relative risk in Port Valdez can be compared to generally accepted and quantitative measures of environmental risk. For these comparisons, we used chemical data collected in Port Valdez. Unless otherwise stated, we chose to exclude data collected before 1992 to prevent complications from upgrades made to the BWTP in 1991. Data available for more than 1 year since 1992 were included to allow for temporal changes. We used two techniques to develop estimates of ecological risk: chemical concentrations in effluent, sediment, and tissues were compared to benchmark values; and the ΣPAH model was used to estimate the acute toxicity to amphipods of ten hydrocarbon compounds in Port Valdez sediments.

Using multiple techniques to evaluate chemical risk provided more than one line of evidence to support the conclusions. A weight-of-evidence approach increases certainty in the risk estimate (Menzie et al. 1996).

Comparison to Benchmark Values

Before using benchmarks, it is important to consider some of the limitations to this approach. Benchmark values were available for some chemicals, but not for other physical or biological stressors that were included in the relative risk estimate. Benchmark values were developed for single compounds and may not account for additive or synergistic effects from multiple stressors. Also, benchmarks may not be low enough for compounds that bioaccumulate (USEPA 1996).

Benchmark values were compared to sediment data that covered five subareas in the Port. Selected benchmarks included ETs established by the EPA and sediment effects range low (ERL) established by the NOAA. If available, marine ET values were chosen for comparison to the Port Valdez sediment data. When ET values were not available, ERL values were used. Table 4.7 describes the metal and hydrocarbon data for sediment collected in 1995 from the small boat harbor (City of Valdez subarea). None of the metals exceeded benchmark values. Several of the PAHs did exceed benchmarks in the six samples that were tested (anthracene, benz[a]anthracene, chrysene, fluoranthene, phenanthrene, and pyrene).

Table 4.7 Sediment Metal and Hydrocarbon Concentrations in 1995 from the Municipal Small Boat Harbor in Port Valdez

Compound in Boat Harbor Sediments	Chosen Benchmark (μg/kg)	Samples Collected in Port Valdez			
		Concentration (μg/kg)	n	Benchmark Exceeded?	Frequency Exceeded
Metals					
Arsenic	8200 (ET-M)	11–18	6	No	0%
Barium	NA	60–103	6	No	0%
Cadmium	1200 (ET-M)	ND– <5	6	No	0%
Chromium	81,000 (ET-M)	55–71	6	No	0%
Mercury	150 (ET-M)	14–21	6	No	0%
Lead	47,000 (ET-M)	ND	6	No	0%
Hydrocarbons					
Anthracene	85.3 (ERL)	0.00–930	6	Yes	16%
Benzo[a]anthracene	261 (ERL)	0.00–754	6	Yes	16%
Chrysene	384 (ERL)	0.00–2630	6	Yes	33%
Fluoranthene	600 (ET-M)	0.00–5610	6	Yes	50%
Phenanthrene	240 (ET-M)	0.50–2980	6	Yes	33%
Pyrene	660 (ET-M)	0.00–3860	6	Yes	33%

Note: ND = below detection limit; NA = not available; ET-M = the marine ET; ERL = sediment effects range low

Table 4.8 contains sediment data collected from 1992 through 1995 from a station adjacent to the BWTP diffuser (Valdez Marine Terminal subarea) and other sampling stations defined by the Alyeska Environmental Monitoring Program (Mineral and Gold Creeks, Anderson Bay, Western Port, and Eastern Port subareas). Samples were also collected at the Ship Escort and Response Vessel System (SERVS) dock in 1992 (City of Valdez subarea) and near the Solomon Gulch Hatchery in 1995 (Dayville Flats and Solomon Gulch subarea). The only compound that exceeded the benchmarks at these locations was 2-methylnaphthalene. This occurred in four of five subareas (Valdez Marine Terminal, Anderson Bay, Western Port, and Eastern Port). Feder and Shaw (1993a) presented gas chromatography–mass spectrometry data that indicated that some of the reported values for 2-methylnaphthalene were overestimates. Because the same analytical method was used between 1992 and 1995, some of these high values for 2-methylnaphthalene may be analytical artifacts.

We identified wildlife benchmarks for the exposure of red fox and mink to benzo[a]pyrene in their diet and compared them to concentrations in mussel tissue sampled by the Alyeska Environmental Monitoring Program from 1992 to 1995, the LTEMP Monitoring Program from 1993 to 1995, and a sea otter disturbance study conducted from 1989 to 1991. The benchmark values were 2.86 and 3.04 mg/kg for red fox and mink, respectively. No exceedences of these benchmarks were observed in any of the subareas sampled (Shoup Bay, Mineral and Gold Creeks, Valdez Marine Terminal, and Sawmill to Seven-Mile Creeks). However, sea otters have a metabolic rate approximately three times that of similar sized terrestrial animals, and the risk to these animals may be underestimated.

Table 4.8   Sediment Hydrocarbon Concentrations from Stations Sampled for Alyeska's Environmental Monitoring Program from 1992 through 1995, Additional Stations Sampled in 1992 at the SERVS Dock Site and in 1995 near the Solomon Gulch Hatchery, and for the LTEMP Monitoring Program from 1995 to 1996

Compound in Port Sediments	Chosen Benchmark (μg/kg)	Samples Collected in Port Valdez			
		Concentration (μg/kg)	n	Benchmark Exceeded?	Frequency Exceeded
Hydrocarbons					
Anthracene	85.3 (ERL)	0.0–19.7	205	No	0%
Acenaphthene	16 (ET-M)	0.0–11.5	205	No	0%
Benzo[a]anthracene	261 (ERL)	0.0–42.2	205	No	0%
Benzo[a]pyrene	400 (ET-F)	0.0–79.0	205	No	0%
Biphenyl	1100 (ET-F)	0.0–12.7	205	No	0%
Chrysene	384 (ERL)	0.0–88.7	205	No	0%
Dibenzo[a,h]anthracene	63.4 (ERL)	0.0–8.7	205	No	0%
Fluoranthene	600 (ET-M)	0.3–105.5	205	No	0%
Fluorene	540 (ET-F)	0.3–13.0	205	No	0%
2-Methylnaphthalene	70 (ERL)	0.0–189.4	205	Yes	4%
Naphthalene	160 (ET-F)	0.0–7.6	205	No	0%
Phenanthrene	240 (ET-M)	0.8–102.6	205	No	0%
Pyrene	660 (ET-M)	0.3–112.1	205	No	0%

Note: Maximum concentrations all occurred in 1992 except for the maximums for biphenyl, dibenzo[a,h]anthracene, and 2-methylnaphthalene, which occurred in 1994. All maximum values are from samples collected at BWTP diffuser station (D33).

ET-F = freshwater ET values; ET-M = marine ET values; ERL = effects range low; ND = below detection limit

We were also able to compare effluent samples from the BWTP to various freshwater or marine benchmarks. All metals tested exceeded the selected benchmark (antimony, arsenic, cadmium, chromium, lead, nickel, thalium, and zinc), with the most exceedences occurring 77% of the time for zinc. Out of the eight PAHs analyzed in the effluent, one (benz[a]anthracene) exceeded the benchmark in 4% of the samples.

Estimating the Risk of Toxicity Due to PAH

Swartz et al. (1995) developed the ΣPAH Model, which applies the concept of additivity to predict the acute toxicity of a mixture of PAHs to marine amphipods. We applied the model to PAH concentrations measured in the sediments of Port Valdez.

The model uses the concentrations of the 13 PAHs from sediment samples collected in the area of interest, predicts the concentration in the sediment pore water, and then predicts the toxicity of these concentrations to amphipods as determined by a large toxicity dataset. Because sediment data from the Port did not consistently include all 13 PAHs used by Swartz et al. (1995), we applied the model to 10 PAHs. Most of the sampling sites were located in the deep offshore areas of the Port (Western Port and Eastern Port subareas), although data were also available from the small boat harbor (City of Valdez subarea) and nearshore areas by Mineral Creek (Mineral and Gold Creeks subarea), the pipeline terminal and BWTP (Valdez

Table 4.9   Acute Toxicity to Amphipods Predicted from Sediment Concentrations of Ten PAHs. The Sum of the Toxic Units, Averaged for All Samples, Is Listed with the Standard Deviations in the Second Column

Subarea and Location	Mean ΣTU ± Std. Dev.	n	Probability of Toxicity	Uncertain Toxicity	No Toxicity
Mineral and Gold Creek	0.001 ± 0.001	24	5%	20%	80%
City of Valdez (Small Boat Harbor)	0.094 ± 0.148	22	5%	20%	80%
Dayville and Solomon Gulch	0.001 ± 0.001	6	5%	20%	80%
* Station nearest the BWTP diffuser	0.011 ± 0.007	12	5%	20%	80%
* BWTP Mixing Zone Stations**	0.002 ± 0.002	75	5%	20%	80%
Western Port	0.001 ± 0.001	36	5%	20%	80%
Eastern Port	0.002 ± 0.002	66	5%	20%	80%

* These stations are located on the border of or in both Subareas G and K.
**Not including Station D33, the station nearest the BWTP diffuser.

Marine Terminal subarea), and the Solomon Gulch Hatchery (Dayville Flats and Solomon Gulch subarea).

The results in Table 4.9 show that none of the acute toxicity levels predicted in Port Valdez occur above the lowest levels set by the model (5% chance of a toxic response). The sum of the toxic units (ΣTU) is included in the table. This value is a measure of the total toxicity associated with environmental concentrations (i.e., field concentration/literature-derived LC_{50}). Although the model indicates little probability of effects in Port Valdez, the ΣTU values are greater in the City of Valdez subarea than in other areas. Sediment samples from this area were collected in the small boat harbor where boat traffic, boat maintenance and repair, and city runoff are likely to contribute to the higher PAH levels in the sediment. The ΣTU was slightly elevated in samples collected at stations on the border between the Valdez Marine Terminal subarea and the deeper eastern Port. These samples were the closest to the discharge of the BWTP.

DISCUSSION

Our effort to assess threats to the highly valued natural resources of Port Valdez, Alaska resulted in the first design and application of a RRM. The regional-scale model established relative risk scores that were generally supported by more traditional and local-scale risk analyses. The relative risk framework accounted for a wide range of chemical (e.g., crude oil and other petroleum products), physical (e.g., vessel traffic and other types of disturbance), and biological (e.g., crowding and gene pool dilution of wild anadromous fishes from hatchery fishes) stressors. For the objectives of this assessment, there were several distinct advantages to considering a regional as well as a site-specific scale. The risk assessment was comprehensive with respect to the array of issues voiced by the stakeholders. Because this comprehensiveness was an essential component of the project, the result was a new and inclusive Port-wide ecological risk perspective. This would not have been possible within the format of a typical EcoRA. Assorted information and data that would typically be disregarded were useful within the context of relative risk. The design of the model allowed many environmental interactions to be considered, documented, and then combined into conclusions that could be revisited as resource managers learned more about this environment. Future utility of the RRM as a risk assessment and resource management tool as well as limitations of the model are discussed below.

Implications of the Relative Risk Model and Confirmatory Analyses

Model results indicated that the areas of most ecological concern were located at the eastern end of the Port, especially at the Duck Flats, an estuarine marsh which was used by a variety of resident, migratory, and over-wintering populations during the year. The near shore environment around the City of Valdez and the BWTP (Valdez Marine Terminal subarea) were also identified as areas at risk of cumulative effects. In general, relative risk was highest for endpoints associated with sediment quality and wildlife food resources. These endpoints tended to be associated with nearshore areas where habitat quality was often high and most of the sources were located. Relative risk was lowest for wild anadromous fishes, bottom fishes, and shellfish. These endpoints tended to be associated with less desirable or diverse habitats and fewer sources. The results of the assessment are reasonable within the context of the Port and suggest that the model demonstrates some degree of accuracy.

We further evaluated the precision of the model and attempted to calibrate the relative risk results by comparing them to risk estimates developed for known chemical concentrations within the Port. The confirmation analyses focused only on chemical stressors and did not address the physical and biological stressors that were included in the RRM. However, enough chemical and toxicity data were available to assess individual as well as multiple chemical stressors. In addition, assessment endpoints related to water quality, sediment quality, and wildlife food resources could be evaluated. The following conclusions could be drawn from the confirmational comparisons:

- Prior amphipod sediment bioassays proved unreliable in the Port due to complicating factors introduced by the fine sediments common in this environment; modeling the sediment toxicity to amphipods circumvented this complication and demonstrated a Port-wide 5% risk of toxicity, the lowest possible result for this model.
- Comparison to ecological benchmarks demonstrated possible risk to receptors from: concentrations of certain PAHs in sediments of the small boat harbor associated with the City of Valdez subarea; the concentration of 2-methylnaphthalene in deep water sediments and along the south shore of the Port; and undiluted concentrations of metals in the treated ballast water discharge.
- Comparison to benchmarks also indicated that there was no chemical risk to wildlife through ingestion of benzo(a)pyrene accumulated in bivalve tissues.

Although the confirmatory analyses suggested that the ecological risk in Port Valdez is low, there is some indication that sediments associated with the small boat harbor (City of Valdez subarea) and the BWTP (Valdez Marine Terminal subarea) exhibit some risk. These results generally agree with the relative risk scores in Table 4.5, where the highest relative risk scores are assigned to the City of Valdez, Duck Flats and Old Valdez, and Valdez Marine Terminal subareas. Sediment samples were not available from Duck Flats and this subarea was not included in the confirmatory analyses.

Field studies conducted in the Port following the completion of this risk assessment continue to provide data that are in general agreement with these risk observations. The LTEMP Monitoring program is collecting intertidal mussel and subtidal sediment samples (Kinnetics Laboratories 2000; Payne et al. 2001). In 2001, samples from the water column, caged mussels, and suspended plastic membrane devices were collected throughout the Port with an emphasis on the BWTP effluent (Payne et al. 2001; Salazar et al. 2002). Hydrocarbons were found in the water throughout the Port at low levels, with slightly elevated levels near the Valdez Marine Terminal and extending out into the eastern Port. Traditional risk assessment techniques indicate that the risk of harm to organisms from these levels remains low (Payne et al. 2001). There is evidence of anthropogenic hydrocarbons near both the Valdez Marine Terminal and the City of Valdez, but some differences in PAH composition have been noted (Kinnetics Laboratories 2000). PAHs were predominantly pyrogenic near the city, while samples collected near the terminal were typical of weathered crude. This difference verifies the presence of a single stressor type with multiple sources in Port Valdez.

Importance of Stakeholder Participation and Scientific Collaboration

The concerns of stakeholders in Port Valdez encouraged us to approach the EcoRA for this area from a new perspective. The primary public concern appeared to center around the most obvious discharger in the Port: the Valdez Marine Terminal. However, when questioned, stakeholders were aware of many other activities or practices that they believed could be affecting their environment. While it was not our intention to reduce concern about the Valdez Marine Terminal, in light of the number of concerns voiced by the public, it was clear that in order to evaluate ecological risk in the Port, it was necessary to consider more than just this one potential source.

This decision was reinforced by the available environmental data, which indicated that contaminant levels were typically at low enough levels that a standard risk assessment would not detect high risk. Based on stakeholder input, the hazard posed by low level but potentially cumulative effects was a concern in this environment where heavy industry and development were mixed with natural resources that were highly valued by the community. The need to address cumulative effects during EcoRAs, as well as the need to simplify complex and interwoven ecological relationships through use of conceptual models and categorization schemes, is well recognized by researchers and natural resource managers working at a watershed scale (Serveiss 2002; Brown et al. 2002; Detenbeck et al. 2000).

An understanding of the dynamics of the ecological system and sound scientific judgment were essential during development of the model. This is especially important in a challenging scenario such as Port Valdez, where a wealth of data was available for specific topics, such as oceanographic measurements and benthic species composition, but little to no information was available about other potentially significant stressors, such as straying hatchery fish. We were able to collaborate with a number of regulators, informed citizens, and scientists who had many years of hands-on experience with the Port environment. However, the final decisions for the model structure and input were completed by the risk assessment team from Western Washington University and researchers from the University of Alaska. The team members from the University of Alaska included Howard Feder and David Shaw. Both had participated in numerous studies and monitoring programs in Port Valdez, as well as the larger Prince William Sound, for more than 20 years. The purpose of many of these studies was to collect baseline ecological data prior to construction of the Valdez Marine Terminal (Feder and Keiser 1980; Feder and Matheke 1980), and then monitoring after construction and during operation (Feder and Shaw 1993a; 1993b; 1994a; 1995; 1996). The close collaboration with these scientists improved the robustness of the model and ensured that important characteristics of this particular ecological system were considered.

Relative Risk Model as a Tool for Risk Assessors and Resource Managers

To effectively evaluate risk within a region, the risk assessor must consider multiple habitats, stressors, and a variety of assessment endpoints. Often data availability is uneven or inconsistent. Interactions and indirect effects among the organisms within the receiving habitat are not well understood. Even with these limitations, the RRM was able to generate relative risk rankings and generate testable hypotheses about the nature of these risks. For example, risk to the Duck Flats is predicted to be greatest from accidental petroleum/chemical spills and physical habitat disturbance resulting from land development and use. Similarly, the shallow subtidal shelf, an area supporting abundant benthic food sources and providing protective cover for fish and wildlife, was identified as a habitat at risk of cumulative effects in the Duck Flats. The model generates these statements by systematically processing an array of information, including scientific data and professional judgment, in a way that is documented and repeatable. The statements can then be tested through additional field investigation and risk estimation. In this manner, the RRM can operate as a guide for risk assessors in Port Valdez. This tool should not reduce the use of direct measurement or quantifiable risk estimates, but should support risk characterization that considers cumulative impacts and draws comparisons that extend beyond a single location.

The model will also serve to inform resource managers. The model indicates that the BWTP discharge was likely to be the major contributor to any effects observed in the offshore water of the eastern Port, but only one of many possible contributors in other areas. Although the magnitude of these effects is not defined, this perspective is useful when applied to future scenarios. If unexplained effects are observed in the eastern Port, the BWTP should be the first source to investigate; however, if impacts were to occur in other areas, other sources should be given equal consideration. The detailed results also point to areas that would significantly reduce risk. For example, in the City of Valdez subarea, accidental spills and contaminated runoff are the largest contributors to risk. Management of these factors would reduce the risk. Better control of vessel traffic and fish waste is not predicted to contribute as much risk reduction in the City of Valdez area.

It is important to note that the model developed in this assessment was based on limited data and should be updated to account for new information, improved understanding of the environment, or more specific concerns. Since this project was completed, various researchers have gained new insights into the distribution and phase composition of the BWTP discharge plume, and have suggested that a surface microlayer may be an important transport factor for hydrocarbons in the Port (Payne et al. 2001; Salazar et al. 2002). Additional toxicity information is also available. Recent toxicity studies have shown that the early life stages of herring and pink salmon are very sensitive to PAHs (Heintz et al. 1999; Carls et al. 1999), which could indicate that habitats where herring or pink salmon spawn are at greater risk from select sources. Redesign and recalculation of the model to account for these factors would increase the sensitivity and educate resource managers about the risk components and exposure pathways in the Port.

Limitations of Relative Risk Models

One of the keys to the utility of the RRM is the detailed listing of assumptions and ranking criteria that have to be made explicit and that can be revisited. But a ranking approach does not provide an estimate of absolute risk. A value of 400 may indicate relatively low risk in some instances, and high risk in others. The model must also be sufficiently thorough and sensitive to predict differences in complex situations, but maintain enough simplicity that the results will reflect some degree of accuracy. If sufficient data are available the values may be calibrated against known dose–response relationships. In the case of Port Valdez, the higher relative risk sites still had relatively low risk when compared to benchmarks and the ΣTU values, suggesting that limited chemical toxicity is occurring in the Port. There was no information to calibrate against other types of stressors, or combinations of stressors.

Introducing greater discriminatory power into the ranking criteria and filter design, or incorporating other data integration techniques may increase the effectiveness of the modeling output. There are several examples of researchers addressing similar problems with multiple stressors and inconsistent data through classification and integration techniques from fuzzy set theory (Harris et al. 1994; Silver 1997).

Another limitation of the relative rank approach is that relative ranks between regions cannot be compared in the same manner as absolute numbers. Two regions, such as Port Valdez and Puget Sound, cannot be compared directly as to low and high risk. Only if the process incorporates both sites with common criteria for ranking sources, habitats, and filters can the two sites be incorporated into a broader risk assessment.

A critical caveat in the interpretation of the ranks and scores of a relative risk assessment approach is the tendency to place the emphasis on the final numerical score or ranking without a careful consideration of the data expressed in tables such as Table 4.5 and Table 4.6. These scores collapse a multivariate construct into just one variable, hiding a great deal of information critical to the decision-making process. The tendency to concentrate on just the summary number is apparent, but should be avoided.

CONCLUSIONS

The RRM is an information management approach combining aspects of EcoRA, mapping, classification, matrix operations, and a weight-of-evidence approach. Regional evaluation allows for the integration of dissimilar information from varied sources and takes advantage of datasets that may not normally be evaluated together. This use of seemingly unrelated data provides comparisons and contrasts that illuminate conditions that may not have been identified prior to the evaluation.

Relative risk rankings in Port Valdez provided a semiquantitative assessment of the cumulative effects that could occur in this watershed-scale ecological system. The assessment was primarily based on professional evaluation of the relative degree to which multiple stressors could interact with multiple receptors and attempted to account for the complexity with which these interactions may occur. By accounting for and systematically evaluating the components of the ecosystem, the RRM formalized a framework for future evaluation. This framework was formed with risk-based principles to encourage system-based decision making during environmental management issues.

REFERENCES

Alyeska Pipeline Service Company. 1995. Mixing Zone Application for NPDES Permit Renewal for the Ballast Water Treatment Plant and Sewage Treatment Plant at the Valdez Marine Terminal Trans Alaska Pipeline System Valdez, Alaska, Anchorage.

Anthony, J.A.M. 1995. Habitat utilization by sea otters (*Enhydra lutris*) in Port Valdez, Prince William Sound, Alaska, M.S. Thesis, University of Alaska Fairbanks.

Brown, B.S., Munns, W.R., Jr., and Paul, J.F. 2002. An approach to integrated ecological assessment of resource condition: the Mid-Atlantic estuaries as a case study, *J. Environ. Manage.*, 66, 411.

Carls, M.G., Rice, S.D., and Hose, J.E. 1999. Sensitivity of fish embryos to weathered crude oil: Part I. Low level exposure during incubation causes malformations, genetic damage, and mortality in larval Pacific herring (*Clupea pallai*), *Environ. Toxicol. Chem.*, 18, 481.

Dambacher, J.M., Li, H.W., and Rossignol, P.A. 2003. Qualitative predictions in model ecosystems, *Ecol. Model.*, 161, 79.

Detenbeck, N.E. et al. 2000. A test of watershed classification systems for ecological risk assessment, *Environ. Toxicol. Chem.*, 19, 1174.

Feder, H.M. and Jewett, S.C. 1988. The subtidal benthos, in *Environmental Studies in Port Valdez, Alaska*, Shaw, D.G. and Hameedi, M.J., Eds., Springer-Verlag, New York, 165.

Feder, H.M. and Keiser, G.E. 1980. Intertidal biology, in *Port Valdez, Alaska: Environmental Studies 1976–1979*, Colonell, J.M., Ed., Institute of Marine Science, University of Alaska Fairbanks, 145.

Feder, H.M. and Matheke, G.E.M. 1980. Subtidal benthos, in *Port Valdez, Alaska: Environmental Studies 1976–1979,* Colonell, J.M., Ed., Institute of Marine Science, University of Alaska Fairbanks, 237.

Feder, H.M. and Shaw, D.G. 1993a. Environmental Studies in Port Valdez, Alaska, 1991 Final Report, Institute of Marine Science, University of Alaska Fairbanks.

Feder, H.M. and Shaw, D.G. 1993b. Environmental Studies in Port Valdez, Alaska, 1992 Final Report, Institute of Marine Science, University of Alaska Fairbanks.

Feder, H.M. and Shaw, D.G. 1994a. Environmental Studies in Port Valdez, Alaska, 1993 Final Report, Institute of Marine Science, University of Alaska Fairbanks.

Feder, H.M. and Shaw, D.G. 1994b. Environmental Survey of the Ship Escort and Response Vessel System (SERVS) Dock Site at Valdez, Alaska, Final Report, Institute of Marine Science, University of Alaska Fairbanks.

Feder, H.M. and Shaw, D.G. 1995. Environmental Studies in Port Valdez, Alaska, 1994 Final Report, Institute of Marine Science, University of Alaska Fairbanks.

Feder, H.M. and Shaw, D.G. 1996. Environmental Studies in Port Valdez, Alaska, 1995 Final Report, Institute of Marine Science, University of Alaska Fairbanks.

Garshelis, D.L. 1983. Ecology of sea otters in Prince William Sound, Alaska, Ph.D. thesis, University of Minnesota, Ann Arbor, MI.

Harris, H.J. et al. 1994. A method for assessing environmental risk: a case study of Green Bay, Lake Michigan, USA, *Environ. Manag.,* 18, 295.

Heintz, R.A., Short, J.W., and Rice, S.C. 1999. Sensitivity of fish embryos to weathered crude oil: Part II. Increased mortality of pink salmon (*Onchorhynchus gorbuscha*) embryos incubating downstream from weathered *Exxon Valdez* crude oil, *Environ. Toxicol. Chem.,* 18, 494.

Kinnetics Laboratories Inc. 1995. Prince William Sound RCAC: Long-Term Environmental Monitoring Program Annual Monitoring Report — 1994, Anchorage.

Kinnetics Laboratories Inc. 1996. Prince William Sound RCAC: Long-Term Environmental Monitoring Program Annual Monitoring Report — 1995, Anchorage.

Kinnetics Laboratories Inc. 2000. Prince William Sound RCAC: Long-Term Environmental Monitoring Program 1999–2000 LTEMP Monitoring Report, Anchorage.

Long, E.R. and Morgan, L.G. 1990. The Potential for Biological Effects of Sediment-sorbed Contaminants Tested in the National Status and Trends Program, NOAA Tech. Memo. NOS OMA 52, U.S. National Oceanic and Atmospheric Administration, Seattle.

Lowell, R.B. et al. 2000. Weight-of-evidence approach for northern river risk assessment: integrating the effects of multiple stressors, *Environ. Toxicol. Chem.,* 19, 1182.

Menzie, C. et al. 1996. A weight-of-evidence approach for evaluating ecological risks: report of the Massachusetts weight-of-evidence workgroup, *Hum. Ecol. Risk Assess.,* 2, 277.

Opresko, D.M., Sample, B.E., and Suter, G.W. 1995. Toxicological Benchmarks for Wildlife: 1995 Revision, ES/ER/TM-86/R2, Lockheed Martin Energy Systems, Inc.

Payne, J.R. et al. 2001. Assessing Transport and Exposure Pathways and Potential Toxicity to Marine Resources in Port Valdez, Alaska, Payne Environmental Consulting, Inc., Encinitas, CA.

Preston, B.L. 2002. Indirect effects in aquatic ecotoxicology: implications for ecological risk assessment, *Environ. Manage.,* 29, 311.

Salazar, M.H. et al. 2002. Final Report 2001 Port Valdez Monitoring, Applied Biomonitoring, Kirkland, WA.

Serveiss, V.B. et al. 2000. Workshop Report on Characterizing Ecological Risk at the Watershed Scale, EPA/600/R-99/111, Office of Research and Development, National Center for Environmental Assessment, Washington, D.C.

Serveiss, V.B. 2002. Applying ecological risk principles to watershed assessment and management, *Environ. Manage.*, 29, 145.

Shaw, D.G. 1996. Unpublished data.

Silver, W. 1997. Ecological impact classification with fuzzy sets, *Ecol. Model.*, 96, 1.

Suter, G.W. 1996. Toxicological benchmarks for screening contaminants of potential concern for effects on freshwater biota, *Environ. Contam. Toxicol.*, 15, 1232.

Swartz, R.C. et al. 1995. ΣPAH: A model to predict the toxicity of polynuclear aromatic hydrocarbon mixtures in field-collected sediments, *Environ. Toxicol. Chem.*, 14, 1977.

U.S. Army Corps of Engineers. 1995. Chemical Data Report Valdez Small Boat Harbor Valdez, Alaska, Valdez.

U.S. Environmental Protection Agency. 1996. Ecotox Thresholds, ECO Update, 3, 1.

U.S. Environmental Protection Agency. 1997. Priorities for Ecological Protection: An Initial List and Discussion Document for EPA, EPA/600/S-97/002, Office of Research and Development, National Center for Environmental Assessment, Washington, D.C.

U.S. Environmental Protection Agency. 1998. Guidelines for Ecological Risk Assessment, EPA/630/R-95/002F, Risk Assessment Forum, Washington, D.C.

U.S. Environmental Protection Agency. 2003. Framework for Cumulative Risk Assessment, EPA/600/P-02/001F, Office of Research and Development, National Center for Environmental Assessment, Washington, D.C.

Wiegers, J.K. et al. 1997. *A Regional Multiple-Stressor Ecological Risk Assessment for Port Valdez, Alaska*, Institute of Environmental Toxicology, Western Washington University, Bellingham.

Wiegers, J.K. et al. 1998. A regional multiple stressor rank-based ecological risk assessment for the fjord of Port Valdez, AK, *Hum. Ecol. Risk Assess.*, 4, 1125.

CHAPTER 5

Application of the Relative Risk Model to the Upper Willamette River and Lower McKenzie River, Oregon

Matthew Luxon and Wayne G. Landis

CONTENTS

1-56670-655-6/04/$0.00+$1.50
© 2004 by CRC Press LLC

This assessment examined risk to spring chinook salmon, rainbow trout, cutthroat trout, and summer steelhead in the mainstem upper Willamette and lower McKenzie Rivers in Oregon, an area of approximately 3,500 km². Regional aspects of the project were addressed using the relative risk model (RRM). The study area was divided into nine risk regions with unique ecological and anthropogenic characteristics. Stressor sources in each region were analyzed and compared to provide a regional perspective of risk.

The RRM ranked risk in each risk region for chemical and physical stressors from multiple sources. The rankings are testable hypotheses regarding the nature and location of risk. Ongoing field studies of periphyton, macroinvertebrate, and fish communities will test these findings.

The results of this assessment show the RRM to be robust for large-scale screening-level ecological risk assessments (EcoRA). Uncertainty in risk predictions was high due to course scale analyses. Site-specific data at the appropriate scale and spatially explicit process models could reduce uncertainty and make the model applicable to higher-tiered risk assessments.

INTRODUCTION

This study is an EcoRA of the upper Willamette River and lower McKenzie River in Oregon. It is designed to determine the relative contribution of natural and anthropogenic stressors to the risk of degradation of the aquatic community and specifically to fish of the family Salmonidae as represented by spring chinook salmon, rainbow trout, cutthroat trout, and summer steelhead. The risk predictions from this assessment are testable hypotheses regarding the area and type of impacts likely to be affecting the ecological structure. The risk analysis was conducted using the RRM developed for an EcoRA of Port Valdez, Alaska (Wiegers et al. 1998).

This project is a component of the National Council of the Pulp and Paper Industry for Air and Stream Improvement (NCASI) Long Term Receiving Water Studies (LTRWS) for pulp and paper mill effluents, an integrated effort to determine the effects of pulp mill effluents on receiving water ecological condition.

The RRM was developed for an EcoRA of Port Valdez, AK (Landis and Wiegers 1997; Wiegers et al. 1998). The RRM follows the risk assessment three-phase approach: problem formulation, analysis, and risk characterization. Each phase is spatially explicit. During the planning phase the location of the project area, habitats supporting potential endpoints, and potential sources of stressors are mapped. In problem formulation the map facilitates discussion between stakeholders, risk managers, and risk assessors to determine assessment endpoints and to break the project area into smaller risk regions with unique combinations of habitats and sources of stress. In the analysis phase a spatially explicit approach ensures that the pathways from release of a stressor to exposure are geographically feasible and accurate. In the risk characterization phase mapping the risk regions provides a means of communicating the nature, location, and extent of the risks of ecological impacts.

Ranks are tied to specific locations within a landscape. The landscape is broken up into risk regions representing unique combinations of sources and habitats. Ranks are assigned based on the distribution of a habitat or source of stressors in a risk region relative to its magnitude in all other risk regions. The numerical ranks for sources and habitats are combined into matrices. The risk scores in the matrices show the degree of overlap between sources of stressors and habitat within each risk region. The matrices thus provide an accounting of risks within a risk region and a comparison of risks among risk regions.

One of the advantages of the relative risk procedure is that it produces testable hypotheses concerning which risk regions are relatively likely to be impacted and the sources contributing to potential impacts. The conceptual model provides mechanistic connections between the measurement endpoints and instream measures of effect. Thus risk predictions should correspond to instream measures of exposure and response relevant to linkages in the conceptual model. Data from the study area that was not used in risk modeling provide tests for these hypotheses.

PROBLEM FORMULATION

Description of the Willamette–McKenzie Study Areas

This risk assessment covers the section of the Willamette River from its inception above Eugene at the confluence with the Coast Fork Willamette River and Middle Fork Willamette River, 56 river miles (RM) down to Corvallis and the McKenzie River from RM 34 to its confluence with the Willamette River near Eugene (Figure 5.1). These boundaries represent the area to be considered in NCASI-LTRWS where they are conducting monitoring of the aquatic ecological structure.

The Willamette River is a ninth-order river, which is unconstrained by topography and flows through the middle of the Willamette Valley. It is the 13th largest river in the United States in terms of discharge (Kammerer, 1990). It flows north for 187 RM from Eugene to Portland, draining the Cascade Mountains to the East, the Calapooya Mountains to the south, and the Coast Fork Mountains to the west. The Willamette Basin is commonly divided into three sections: the upper, middle, and lower basins (Willamette Basin Task Force 1969). This study focuses on the upper basin, which includes the river from Eugene to Corvallis, RM 187 to RM 128. At Corvallis, the Willamette River drains approximately 4000 square miles including the watersheds of the Coast Fork Willamette River, the Middle Fork Willamette River, the Long Tom River, the McKenzie River, and the Mary's River. At RM 161 the average discharge for water years 1969 to 1998 is 11,490 cfs (USGS 1999).

The McKenzie River is a seventh-order river. It flows mainly west 93 miles through the Cascade Mountains. Major tributaries include the Blue River, the South Fork McKenzie River, and the Mohawk River. The McKenzie Basin drains approximately

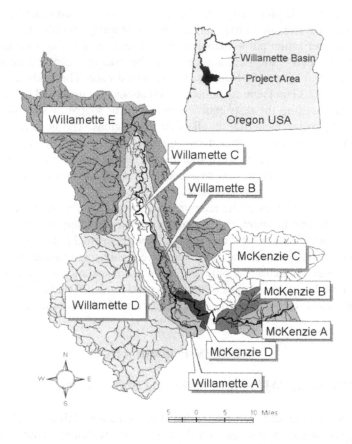

Figure 5.1   Risk regions of the project area and its location within the Willamette Valley.

1300 square miles where it joins the Willamette River at RM 174, approximately doubling the flow of the Willamette River. Average discharge at RM 6.5 is 5933 cfs. This study focuses on the lower 34 miles of the McKenzie River from Dearhorn Park, where the valley widens into a broad alluvial floodplain and the river becomes unconstrained, to the confluence with the Willamette River.

Assessment Endpoints

To best represent the desired state of the ecological structure the assessment endpoints must have social and biological relevance, be accessible to prediction and measurement, and be susceptible to the hazard being assessed (Suter 1990). For this EcoRA, social relevance is assured by deriving endpoints from stakeholder concerns. Supporting the stated values with numerical criteria from state regulations ensures that endpoints for this EcoRA are accessible to prediction and measurement. Criteria to establish assessment endpoints are derived from values expressed through the Willamette Valley Livability Forum (WVLF) (WVLF 1999) and the Willamette

Basin Reservoir Study (WBRS) (OWRD 1999a). State regulations provide quantitative criteria related to the expressed values (Table 5.1).

Although there are other values expressed through the WVLF and WRBS studies, this assessment focuses on the aquatic environment and the directly related values. Maintenance of self-sustaining populations of native salmonids and providing a recreational and commercial fishery for salmonids are the primary values evaluated in this risk assessment. Water quality criteria for the support of aquatic life are used because these criteria reflect physical conditions and concentrations of toxicants that, through testing and extrapolation, are believed to depress populations of valued species.

Stakeholder values for the Willamette Basin extend beyond the native salmonids to include such uses as swimming; protection from flooding; providing sources of water for drinking, irrigation, and industry; maintenance of reservoirs for windsurfing and warm water fisheries; and providing a fishery for introduced salmonids such as coho and summer steelhead. These values may directly conflict with the stated value of maintaining a self-sustaining native fish population. Inclusion of these values in the relative risk assessment enables identification of areas where compromise in maintaining the resources may be necessary.

Receptors of Concern

There are 28 species of fish reported in the project area (Altman et al. 1997), of which eight are salmonids. Five species of fish that use the project area at least occasionally, including river lamprey, pacific lamprey, spring chinook salmon, Oregon chub, and bull trout, are listed under the Endangered Species Act. The salmonids spring chinook salmon, cutthroat trout, rainbow trout, and summer steelhead are selected as receptors of concern because they spawn within the project area and are highly valued as sport and food fish. Habitat requirements for the salmonids selected as receptors of concern (ROCs) overlap to a large degree with those of the other listed fish, thus assessing risk for these salmonids to some degree addresses risk to the other listed fish.

Sources of Stressors

The United States Geological Survey (USGS) recently included the Willamette Basin in their National Water Quality Assessment showing that it is moderately degraded compared with other rivers in the United States of similar size. They showed that fish community structure is degraded except in higher elevation streams. More pollution-tolerant species and more external anomalies were found in streams with few riffles, poor riparian habitat, and elevated temperatures. Water chemistry was not strongly associated with fish community structure except that external anomalies and pollution-tolerant fish were associated with agricultural and urban streams with the highest nutrient and pesticide concentrations. Additionally, they show that nutrients in streams and groundwater are degrading water quality (Wentz et al. 1998).

Table 5.1   Stakeholder Values and Assessment Endpoints

Stakeholder Values	Assessment Endpoint	Citation
Water Quality		
River water is usable as a source of drinking water	River water meets or exceeds Oregon drinking water quality criteria	1
River is swimmable	River water meets or exceeds Oregon water quality criteria for primary contact recreation	1
Avoiding or minimizing point and nonpoint sources of pollution (chemical input into the river does not compromise water quality)	Water meets or exceeds Oregon water quality criteria for toxics and nutrients	1
Conservation is the primary means of ensuring an adequate supply of water	Management by wise utilization of the water supply innate to the watershed	1
Fish caught from the river are palatable and safe to eat	River meets or exceeds the Oregon water quality criteria for aquatic life	1
Fisheries		
There are sufficient numbers of desirable fish to support an active recreational and commercial fishery	No loss of fish production	
	No reduction of allowable catch of sport fish	
Summer steelhead	Population meets ODFW[a] basin fisheries plan: maintain a potential sport catch of 250 in the mainstem above Willamette Falls	6
	Maintain an annual catch of 1200 on the McKenzie River	3
	Maintain a return of 2400 to the McKenzie sub-basin	3
Native Fish Populations		
River sustains thriving populations of native fish	Populations of spring chinook salmon, rainbow trout, cutthroat trout, and winter steelhead meet ODFW basin fishery plans	2–5
Spring chinook salmon	Increase production to 100,000 fish entering the Columbia River	2
	Increase the number of wild spring chinook salmon to the McKenzie River to 10,000	2
Rainbow trout	No hatchery rainbow trout found below Hayden Bridge in the McKenzie River	3
	No detectable loss of current production	3,5
Cutthroat trout	No detectable loss of current production	3,5
Winter steelhead	Maintain current annual sport catch in the upper Willamette River of 190 fish	3,4
	No loss of genetic diversity of native salmonid species	2–5
Habitat		
Floodplain protection and enhancement for natural functions and values	No net loss of riparian or floodplain vegetation	

Table 5.1   Stakeholder Values and Assessment Endpoints (continued)

Habitat

Floodplain protection and enhancement for natural functions and values	No net loss of riparian or floodplain vegetation

Potentially Conflicting Values

Floodplain management for human health and safety	Flow is controlled to prevent damage to human lives or property in urban areas
Water quantities sustain human communities	Crop irrigation
	Human consumption
Maintaining reservoirs for fishing, boating and windsurfing	
No loss of recreation including	
Boating	
Fishing	e.g., summer steelhead

Values expressed through: Willamette Basin Reservoir Study (OWRD 1999a)

Citations:

1) Oregon Administrative rules, chapter 340, Division 41, 1994
2) ODFW 1998
3) ODFW 1991
4) ODFW 1990
5) ODFW 1988

ª Oregon Department of Fish and Wildlife.

The Willamette River from its mouth to RM 190 was listed under Section 303d of the Clean Water Act as water quality impaired for toxics due to elevated mercury levels in fish (ODEQ 1998). In 1997 the Oregon Health division issued a mercury advisory for consumption of smallmouth bass, largemouth bass, and northern squawfish from the entire mainstem Willamette River. Both the Willamette River and the McKenzie River are listed under Section 303d as water quality impaired for temperature due to elevated summer temperatures. The indigenous populations of spring chinook salmon, Oregon chub, and bull trout have declined to the point that they are listed as federally threatened or endangered under the Endangered Species Act.

The sources of stressors being considered in this assessment are forestry, urbanization, agriculture, water withdrawals, industrial effluent, introduced hatchery fish, and habitat alteration (Table 5.2).

Risk Regions

The project area was divided into nine risk regions (Table 5.3 and Figure 5.1) based on the distribution of salmonid spawning, rearing, and migration habitat and on the initial assessment of the distribution of sources of stressors. The adjacent subwatersheds draining to the specified sections of river delineate the boundaries of the risk regions. Relative risk analyses compare the relative magnitude of sources of stress occurring in and affecting habitats contained within these smaller watersheds.

Table 5.2   Sources of Stressors and Examples of Stressors Released

Source of Stressors	Occurrence in and Upstream of Project Area	Associated Stressors
Forestry	About 90% of the lands draining to the project area are forested and the majority of these lands are managed for timber harvest	Increased sediments Catastrophic debris flows Increased temperatures
Urbanization	Ten population centers including Eugene–Springfield (combined population 190,000) and Corvallis (population 52,000)	Increased sediments and nutrients Metals and organic pollutants including pesticides, industrial and automotive wastes Increased peak flows and lower low flows
Agriculture	41% of project area in agriculture (primarily grass seed, mint, and filberts)	Pesticide runoff Increased sediments and nutrients
Industrial effluent	317 permitted waste dischargers in the project area and 22 upstream of the project area	Metals, organics, BOD,[a] TSS,[b] and nutrient input Increased temperature
Water withdrawal	Two run-of-the-river dams, multiple industrial and agricultural water rights, numerous unpermitted withdrawals	Flow-related reduction in spawning and rearing habitat Flow-related increases in temperatures
Hatchery fish	Four hatcheries that release rainbow trout, summer steelhead salmon, and spring chinook salmon into or above the project area	Hybridization and competition with wild fish
Habitat alteration	75% of historic shoreline lost	Changes in current patterns Loss of refugia Loss of instream cover Loss of bank cover

[a] BOD, biological oxygen demand.
[b] TSS, total suspended solids.

Risk Characterization

Several analyses were conducted to determine the extent of the above stressors and habitats in each risk region. The analyses are not exhaustive, but provide a measure of the magnitude of a given source of stressors within each risk region relative to other risk regions within the project area. Analyses measure the magnitude of a given stressor within each risk region or within the cumulative watershed contributing to the indicated risk region. A cumulative watershed is the risk region plus all land upstream of the risk region. An example of a cumulative watershed is that of Willamette (WB). The cumulative watershed contributing to WB includes the entire McKenzie Basin, the upstream risk regions McKenzie A (MA), and the Middle Fork and Coast Fork Willamette watersheds above MA. Ranks show the magnitude of the source or habitat in a given risk region relative to the other risk regions in the project area.

Each source and habitat was ranked for each risk region to indicate high, moderate, low, or no magnitude. Ranks are assigned using criteria specific to the project area. Criteria are based on the size and frequency of the source and the amount and quality of available habitat. Ranks are assigned to each source and habitat type on a two-point scale from 0 to 6 where 0 indicates lowest magnitude and 6 the highest.

Table 5.3 Features of the Designated Risk Regions

Risk Region	River Miles	Physical Features	Spawning Habitat	Rearing Habitat	Migration Habitat	Stressors
McKenzie A (MA)	34.5 to 21.7	Floodplain: moderately constrained Sinuousity: 1.16 Slope: 0.4% Bed: cobble/gravel	Spring chinook Rainbow trout	Spring chinook Rainbow trout Cutthroat trout Summer steelhead	Spring chinook Rainbow trout Cutthroat trout Summer steelhead	Walterville diversion canal Forest landuse Upstream hatchery
McKenzie B (MB)	20.7 to 16	Floodplain: wide unconstrained Sinuousity: 1.12 Slope: 0.4% Bed: cobble/gravel	Spring chinook Rainbow trout	Spring chinook Rainbow trout Cutthroat trout Summer steelhead	Spring chinook Rainbow trout Cutthroat trout Summer steelhead	Urban, agricultural, and forest landuse Weyerhaeuser pulp and paper mill (NPDES major industrial discharger)
McKenzie C (MC)	16 to 13	Floodplain: unconstrained Sinuousity: 1.05 Slope: 0.2% Bed: cobble/gravel	Spring chinook	Spring chinook Rainbow trout Cutthroat trout Summer steelhead	Spring chinook Rainbow trout Cutthroat trout Summer steelhead	Outskirts of Springfield Mohawk River (forest and agricultural landuse) Weyerhaeuser pulp and paper mill 18 permitted dischargers
McKenzie D (MD)	13 to 0	Floodplain: unconstrained Sinuousity: 1.08 Slope: 0.2% Bed: cobble/gravel	Rainbow trout	Spring chinook Rainbow trout Cutthroat trout Summer steelhead	Spring chinook Rainbow trout Cutthroat trout Summer steelhead	Outskirts of Eugene–Springfield on the left bank and agricultural and urban land on the right bank 13 NPDES dischargers

Table 5.3   Features of the Designated Risk Regions (continued)

Risk Region	River Miles	Physical Features	Spawning Habitat	Rearing Habitat	Migration Habitat	Stressors
Willamette A (WA)	186 to 175	Floodplain: unconstrained Sinuousity: 1.08 Slope: 0.19% Bed: cobble/gravel Depth: shallow	Rainbow trout	Spring chinook Rainbow trout Cutthroat trout Summer steelhead	Spring chinook Rainbow trout Cutthroat trout Summer steelhead	Eugene–Springfield Many small permitted dischargers Two sewage treatment plants
Willamette B (WB)	175 to 160	Floodplain: unconstrained Sinuousity: 1.08 Slope: 0.19% Bed: cobble/gravel Depth: shallow		Rainbow trout Cutthroat trout	Spring chinook Rainbow trout Cutthroat trout Summer steelhead	Agricultural landuse City of Harrisburg runoff and STP
Willamette C (WC)	160 to 149	Floodplain: unconstrained Sinuousity: 1.13 Slope: 0.08% Bed: cobble/gravel/sand Depth: deep		Rainbow trout Cutthroat trout	Spring chinook Rainbow trout Cutthroat trout Summer steelhead	Agricultural landuse Small sewage treatment plant on tributary
Willamette D (WD)	149 to 143	Floodplain: unconstrained Sinuousity: 1.18 Slope: 0.08% Bed: cobble/gravel/sand Depth: shallow Braided channel			Spring chinook Rainbow trout Cutthroat trout Summer steelhead	Urban and agricultural landuse Industrial waste discharge (Amazon Creek) Long Tom River Two pulp and paper mills
Willamette E (WE)	134 to 128	Floodplain: unconstrained Sinuousity: 1.21 Slope: 0.01% Bed: cobble/gravel Depth: shallow			Spring chinook Rainbow trout Cutthroat trout Summer steelhead	Corvallis Mary's River Muddy River Agriculture, forestry, and urban landuse

Note: Sinuosity: actual channel distance divided by straight-line distance
NPDES: National pollution discharge elimination system

Table 5.4 Habitat Quality Ranks for Freshwater Lifestages of ROCs in the Upper Willamette Basin

	Spawning	Rearing	Migration	Sum	Spawning	Rearing	Migration	Sum	Spawning	Rearing	Migration	Sum	Spawning	Rearing	Migration	Sum
	McKenzie A				McKenzie B				McKenzie C				McKenzie D			
Spring Chinook	4	6	6	16	2	6	6	14	0	6	6	12	0	6	6	12
Summer Steelhead	0	4	6	10	0	2	6	8	0	2	6	8	0	2	6	8
Rainbow Trout	6	6	6	18	6	6	6	18	6	6	6	18	6	6	6	18
Cutthroat Trout	0	4	6	10	0	4	6	10	0	6	6	12	0	6	6	12
Sum	10	20	24	54	8	18	24	50	6	20	24	50	6	20	24	50

	Spawning	Rearing	Migration	Sum	Spawning	Rearing	Migration	Sum	Spawning	Rearing	Migration	Sum	Spawning	Rearing	Migration	Sum	Spawning	Rearing	Migration	Sum
	Willamette A				Willamette B				Willamette C				Willamette D				Willamette E			
Spring Chinook	0	2	6	8	0	6	6	12	0	6	6	12	0	4	6	10	0	2	6	8
Summer Steelhead	0	0	6	6	0	0	6	6	0	0	6	6	0	0	6	6	0	0	6	6
Rainbow Trout	4	4	4	12	2	4	4	10	0	2	2	4	0	0	0	0	0	0	0	0
Cutthroat Trout	0	4	6	10	0	6	6	12	0	4	4	8	0	2	2	4	0	2	2	4
Sum	4	10	22	36	2	16	22	40	0	12	18	30	0	6	14	20	0	4	14	18

Habitat

The quality of the instream habitat of the mainstem Willamette River and mainstem McKenzie River for the ROCs was determined for each of the risk regions subjectively through discussion with an ODFW fish biologist (Wade 1999a). Ranks indicate the capacity of the mainstem river segment of each risk region to support spawning, rearing, and migration lifestages for the various fish species used as assessment endpoints. It is assumed that habitats supporting salmonids also support other valued ecological attributes such as water quality. Overall high ranks for a risk region thus represent the ability of that risk region to support water quality values for aquatic life beyond the salmonids indicated. Table 5.4 shows the results of the habitat ranking. The general trend is increasing habitat quality and more life stages supported proceeding from the downstream risk regions to upstream risk regions (i.e., from Willamette E [WE] to MA and WB).

Sources of Stressors

Ranks were assigned to all sources of stressors except hatchery fish based on the statistical distribution of sources of stressors among the subareas. This eliminates bias associated with subjective assignation of ranks. Jenk's optimization statistical analysis (Groop 980) using ArcView™ 3.1 software was used to assign ranks. Jenk's optimization determines natural breakpoints in data by ordering the data from low to high and interactively assigning groups by minimizing the variance within groups and maximizing the variance between groups until an optimal grouping is determined for the number of groups the user specifies. In general, data for each source were assigned to four natural groups. Ranks of 0, 2, 4, and 6 were assigned to these groups from lowest to highest, respectively, to indicate no, low, medium, and high magnitude of a given source in the indicated risk region.

Table 5.5   Percent Area in Agricultural Landuse, Urban Roads, and Forest Roads for Risk Region Cumulative Watersheds

Risk Region	% Area in Agriculture	Rank	% Area in Urban Roads	Rank	% Area in Forest Roads	Rank
MA	0.8	0	0.3	0	1.32	4
MB	0.9	0	1.7	2	1.28	4
MC	1.0	0	0.3	0	1.43	6
MD	1.1	0	4.1	4	1.40	6
WA	3.4	4	13.5	6	0.87	0
WB	2.0	2	1.7	2	1.09	2
WC	2.5	2	0.7	0	1.05	2
WD	5.0	4	1.0	2	0.98	2
WE	6.9	6	0.6	0	0.97	2

Landuse

Cumulative Landuse

Landuse provides an estimate of the types of activities occurring in the watershed that may adversely affect river ecology. The proportion of antropogenic landuses (urban, agriculture, or forestry) for the cumulative watershed was determined.

Agricultural landuse was calculated as a straightforward proportion of overall landuse, whereas the magnitude of forestry and urban landuse in each risk region was determined slightly differently. For these sources of stressors, the proportion of each watershed occupied by forest roads or urban roads* was calculated. This analysis assumes that watersheds with more roads in forest landuse, for example, have more logging activity and thus contribute more stress associated with logging and the infrastructure supporting it. Urban roads provide a surrogate for impermeable surfaces, which through the resultant change in the hydrologic regime have been shown to be a leading factor in degrading physical habitat in urban watersheds. Results reflect the relatively high proportion of forestry in the upper Willamette and McKenzie watersheds, increasing agriculture in the lower portions of the watershed, and the influence of urbanization in the Eugene–Springfield area (Table 5.5).

Riparian Landuse

Anthropogenic landuse (agriculture and urban) in the riparian zone was evaluated. Landuse in the riparian zone directly affects habitat quality in the adjacent river by determining refugia during flood events, large woody debris recruitment potential, and shade. The riparian zone is defined in this case as the land flooded during the February 1996 flood, which was of a 30-year recurrence interval.

Analyses were carried out by overlaying geographic information system (GIS) coverage of the 1996 flood (ACOE 1998) with the aforementioned landuse GIS

---

* Urban landuse was calculated as a proportion of the risk region, not the cumulative watershed.

Table 5.6   Percent of 30-Year Floodplain in Agricultural or Urban Landuse

Risk Region	% Agriculture	% Urban	Total % Anthropogenic	Rank
MA	57	0	57	4
MB	23.7	1.3	25	2
MC	18.6	1.3	19.9	0
MD	36.4	1.3	37.7	2
WA	9.4	27.4	36.8	2
WB	72.7	0.1	72.8	6
WC	61.9	0	61.9	4
WD	25.6	1.1	26.7	2
WE	68.5	0.9	69.4	6

coverages to determine percent of anthropogenic landuse within the floodplain for each risk region.

Results presented in Table 5.6 show that risk regions WB and WE had relatively large amounts of agriculture occurring close to the river. The analysis also shows that agriculture is largely confined to the riparian zone along the McKenzie River. Urbanization in the riparian zone is largely confined to risk region WA, but some urban riparian influence occurs in most risk regions.

Water Rights

Impacts due to water withdrawal were calculated by determining the cumulative water rights for agricultural and industrial water withdrawal. Water rights represent the maximum amount of water that a permit holder may use. This analysis assumes that the proportion of a given water right used to that allotted is similar among risk regions. The sum of water rights for the cumulative risk region was calculated based on the Oregon Water Resources Department's points-of-diversion GIS and water rights information system database (OWRD 1999b). Total water rights for a given risk region (including upstream areas) were normalized to the modeled July-to-September low flow rate at the bottom of each risk region (Laenen and Risley 1997). This indicates that a large withdrawal is more important in an area with low flow than an area with high flow.

Water rights are awarded for storage of water as a volume (acre-foot [ac-ft]), while surface water and groundwater rights are awarded as a rate (cubic feet per second [cfs]). Ranks for the relative magnitude water withdrawals among risk regions were determined by assigning ranks to both stored and surface water withdrawals separately. Stored and surface water ranks were then averaged and rounded up to the nearest even number to determine the final rank.

Results show that the permitted withdrawal rate from MA is much larger than that for other risk regions (Table 5.7). This withdrawal represents the Walterville diversion canal. Water is withdrawn for the length of MA to power turbines for electricity production. The water is returned to the river at the inception of risk region MB, thus it is not reflected in rank scores for downstream risk regions. Large agricultural reservoirs in WD and WE represent the largest stored water rights for all risk regions.

Table 5.7 Ratio of Water Rights to Predicted River Low Flow

Risk Region	CFS		Acre-Ft		
	Percent	Rank	Ratio	Rank	Avg. Rank
MA	42	6	0.15	0	4
MB	3	2	0.15	0	2
MC	9	2	0.44	2	2
MD	9	2	0.43	2	2
WA	22	4	0.76	4	4
WB	17	4	0.34	2	4
WC	19	4	0.34	2	4
WD	33	6	2.65	6	6
WE	36	6	2.74	6	6

Table 5.8 Ratio of NPDES Permitted Waste Dischargers Permitted Effluent Flow to Predicted River Low Flow

Risk Region	Effluent Flow (cfs)	River Flow (cfs)	Ratio	Rank
MA	0.77	2500	0.000	0
MB	0.77	2500	0.000	0
MC	46.98	2540	0.018	2
MD	46.98	2600	0.018	2
WA	56.4	2030	0.028	6
WB	104.51	4640	0.023	4
WC	109.76	4750	0.023	4
WD	132.8	4940	0.027	6
WE	144.8	5090	0.028	6

Industrial Effluent

The magnitude of industrial effluent discharge was based on the volume of effluent contributed from dischargers within and upstream of each risk region based on USEPA industrial facilities discharge data from 1990 to 1996 for NPDES minors flow data (EPA 1998a) and EPA Toxic Release Inventory data from 1993 to 1999 for NPDES majors flow data (EPA 1999). To account for differing dilution capacities of the river in the various risk regions, total effluents were normalized to instream flow at the downstream end of each risk region based on modeled streamflow data (Laenen and Risley 1997). Low flow estimates and NPDES majors effluent data for July, August, and September were used to arrive at the lowest dilution ratio. Yearly average effluent flow data for NPDES minors were also added to the NPDES majors data, which increased the volume of effluent in each risk region by a small amount beyond that calculated from NPDES majors summer flow volume alone. Results (Table 5.8) show the ratio of effluent to stream volume tends to increase in a downstream fashion despite increased stream flow downstream.

Table 5.9   Ratio of Length of Revetments to
Length of Mainstem River Segment

Risk Region	Ratio	Rank
MA	0.29	4
MB	0.32	4
MC	0.17	0
MD	0.24	2
WA	0.25	2
WB	0.50	6
WC	0.33	4
WD	0.37	4
WE	0.24	2

Channel Modification

The amount of riverbanks stabilized with revetments was used as an indicator of the degree to which the river channel has been modified. Army Corps revetments GIS data (ACOE 1999a, ACOE 1999b) were used to calculate the linear distance of revetments in risk regions as a proportion of linear distance of mainstem McKenzie or Willamette River. Because revetments can occur on both banks, the percentage of river covered by revetments could range up to 200%.

Results show that revetments occur in all risk regions (Table 5.9). However, risk region WB, with half of the river miles in revetment, had by far the most.

Hatchery Releases

ODFW hatchery release data (Wade 1999b) were used to determine the number of fish released from hatcheries in the vicinity of each risk region. Hatcheries release rainbow trout, spring chinook salmon, and summer steelhead into the McKenzie River and the Willamette River. Additionally, winter steelhead are released into the Middle Fork Willamette River. The anadromous fish are released from the hatcheries, and some presmolts are released in other areas in an attempt to establish naturalized runs. Legal size rainbow trout are released from boats and spread evenly along segments of the McKenzie River above Hayden Bridge and in the Middle Fork Willamette River and Coast Fork Willamette River.

Ranks were assigned assuming that the density of hatchery fish is greatest near the release site and that competition with native spawners occurs only in risk regions where hatchery fish are released (for spring chinook salmon and rainbow trout). The numbers of hatchery fish competing with natives for spawning sites were assumed to be greater closer to release sites than further away.

A rank of 6 was assigned to risk regions where hatchery fish are released. A rank of 4 was assigned to neighboring risk regions and a rank of 2 to all other risk regions where spawning or rearing of native fish occurs (Table 5.10).

Table 5.10 Habitat Ranks Used with RRM

Risk Region	Spring Chinook Salmon	Rainbow Trout	Cutthroat Trout	Summer Steelhead	Total
MA	16	18	10	10	54
MB	14	18	10	8	50
MC	12	18	12	8	50
MD	12	18	12	8	50
WA	8	12	10	6	36
WB	12	10	12	6	40
WC	12	4	8	6	30
WD	10	0	4	6	20
WE	8	0	4	6	18
Sum	104	98	82	64	

Table 5.11 Source of Stressor Ranks Used with RRM

Risk region	Cumulative agriculture	Riparian development	Logging	Urbanization	Water withdrawal	Revetments	NPDES	Hatchery	Total
MA	0	4	4	0	4	4	0	6	24
MB	0	2	4	2	2	4	0	4	16
MC	0	0	6	0	2	0	2	2	16
MD	0	2	6	4	2	2	2	2	16
WA	4	2	0	6	4	2	6	4	30
WB	2	6	2	2	4	6	4	2	22
WC	2	4	2	0	4	4	4	2	20
WD	4	2	2	2	6	4	6	2	26
WE	6	6	2	0	6	2	6	2	30
Sum	18	26	28	6	34	28	20	26	

Summary of Sources and Habitats of Stressor Ranks

Table 5.10 and Table 5.11 show the final ranks for all sources of habitats and stressors, respectively. These data provided the basic input to the RRM.

Calculation of Relative Risks

Calculation of relative risk occurs in two steps. First, the source and habitat ranks are multiplied to calculate a preliminary relative risk score (Table 5.12). Second, exposure filters are applied based on the likelihood a source of stressors releasing stressors to a habitat (Landis and Wiegers 1997) i.e., through application of best

Table 5.12   Relative Risk Matrix: Sources of Stressors

Risk Region	Cumulative agriculture	Riparian development	Logging	Urbanization	Water withdrawal	Revetments	NPDES	Hatchery	Sum	NPDES	Hatchery	Sum
	Filtered Relative Risk Scores									Unfiltered Relative Risk Scores		
MA	0	216	216	0	216	216	0	324	1188	0	264	1128
MB	0	100	200	100	100	200	0	200	900	0	168	868
MC	0	0	300	0	100	0	100	100	600	52	84	536
MD	0	100	300	200	100	100	100	100	1000	52	84	936
WA	144	72	0	216	144	72	216	144	1008	84	120	852
WB	80	240	80	80	160	240	160	80	1120	72	68	1020
WC	60	120	60	0	120	120	120	60	660	48	48	576
WD	80	40	40	40	120	80	120	40	560	36	28	464
WE	108	108	36	0	108	36	108	36	540	24	24	444
SUM	472	996	1232	636	1168	1064	924	1084		368	888	

professional judgment of sources not likely to release a stressor to a habitat are eliminated from relative risk estimates by multiplying the resulting risk estimates by zero.

Two source–habitat combinations were determined to require exposure filters of 0 indicating low likelihood of a complete exposure pathway: hatchery fish and permitted discharges. All other sources of stressor–habitat combinations received exposure filters of 1, indicating a complete exposure pathway. Permitted discharges are likely to have little effect on migration behavior of the ROCs because the fish migrate during high flows when dilution is greatest. Thus, a filter of 0 was applied to migration habitat for the source term industrial effluent. Hatchery-reared fish are likely to have no undesirable effects on summer steelhead since summer steelhead are themselves a hatchery-reared fish, and natural reproduction is not desirable in the project area. Thus, a filter of 0 was applied to summer steelhead for the source of stressors hatchery fish.

RRM RESULTS

Relative risk scores for sources and habitats are presented in Table 5.12 and Table 5.13, respectively, and in Figure 5.2. Scores for each risk region are determined by summing across each row in the matrices. High relative risk scores indicate relatively high habitat value overlapping with relatively high magnitude sources of stressors.

Table 5.13 Relative Risk Matrix: Habitat

Risk Region	Spring Chinook Salmon	Summer Steelhead	Rainbow Trout	Cutthroat Trout	Sum
MA	352	396	220	160	1128
MB	252	324	180	112	868
MC	132	204	132	68	536
MD	228	348	228	132	936
WA	188	312	244	108	852
WB	312	264	312	132	1020
WC	240	80	160	96	576
WD	244	0	100	120	464
WE	204	0	108	132	444
Sum	2152	1928	1684	1060	

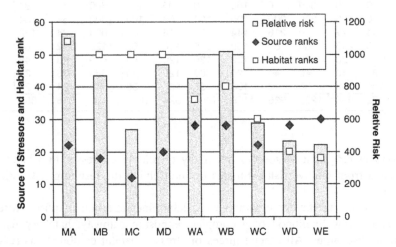

Figure 5.2 Cumulative habitat and source ranks input into the RRM (left axis) and cumulative risk scores output from the model (right axis).

Figure 5.2 shows how habitat and source ranks interact across the project area to produce the overall risk ranks. The general pattern shows that relatively high habitat ranks drive the risk in the McKenzie River risk regions, whereas relatively high source ranks drive the risk in the Willamette River risk regions. For example, although high-magnitude sources are likely to release stressors to risk regions WD and WE, there is a relatively low risk of their interacting with ROCs. Similarly, although there is an abundance of high-quality habitat in risk region MC, there is a relatively low risk of stressors released to that habitat.

Relative risk scores for sources of stressors that are relatively high over the entire project area (the bottom row of Table 5.12) are logging, water withdrawal, and revetments. These sources of stressors suggest that habitat loss and degradation contribute the greatest risk to ROCs in the project area. The high relative risk score for hatchery fish indicates that risks associated with habitat loss and degradation are likely to be compounded by risk from competition. In addition to these stressors, the project area as a whole has significant contributions from all sources of stressors; thus, a variety of stressors may interact to affect the ROCs. By examining the body of the sources of stressors and habitat matrices we are able to predict which of the sources of stressors are likely to interact with the various ROCs.

By examining the component source of stressor scores in the body of the source of stressors matrix (Table 5.12) it is possible to predict stressor interactions in the risk regions. For example, in the McKenzie Basin, risk region MA has high component relative risk scores for riparian development, logging, and water withdrawal. These sources of stressors release stressors that directly affect habitat quality and abundance and suggest high risk from habitat degradation. These scores, combined with the high relative risk score for hatchery fish, suggest that competition with non-native fish may interact with habitat loss in MA, compounding risks to the ROCs. The primary sources of stressors that release toxic stressors — urbanization, agriculture, and NPDES — are relatively low in MA, suggesting that toxicity risks are low in MA relative to other risk regions. MB is similar to MA, except that both the habitat and source of stressors ranks are lower. Thus, although the stressors are similar, the relative risk is lower. In contrast, although the overall relative risk for MC is low, there is a high risk from logging associated with the heavily logged Mohawk watershed. Thus, stressors associated with logging are likely to interact with ROCs in MC, but they are relatively unlikely to be compounded by stressors from other sources in comparison to the other McKenzie River risk regions.

The patterns of sources of stressors that contribute to relative risk scores for the Willamette River risk regions are somewhat different from those of McKenzie River risk regions. Where the largest contributor to relative risk in the McKenzie basin is logging, logging contributes comparatively little to risks in the Willamette River risk regions. In the Willamette River risk regions, there are relatively greater contributions from sources of stressors that release toxic stressors. Because the source of stressors scores for revetments and riparian development are also high in the Willamette risk regions, toxic stressors in the Willamette River risk regions may compound risks to ROCs from habitat degradation. Looking at the high-ranking Willamette River risk regions, the overall risk score for WB is similar to that of MA; however, WB has some contribution from all of the sources of stressors. Thus, in WB, a wide variety of stressors is likely to compound the risks associated with habitat degradation-related stressors released by the highest-ranking sources, revetments, and riparian development. In WA, high-quality habitat overlaps with intense urbanization associated with the Eugene–Springfield area, as the Willamette River at WA demarcates the boundary between these two cities. Accordingly, relative risk scores are largely driven by sources of stressors indicating urbanization (NPDES and urban landuse) with additional substantial contributions from all other sources of stressors except logging.

The relative risk scores for each habitat over the entire project area are shown in the bottom row of Table 5.13. Relative risk by habitat shows that native species (spring chinook salmon, rainbow trout, and cutthroat trout), which use the river for a greater proportion of their lives, have the highest overall risk scores. By examining the body of the risk matrix (Table 5.13) for each species by risk region, we can see which risk regions are contributing the greatest risk to each species. Thus, we see

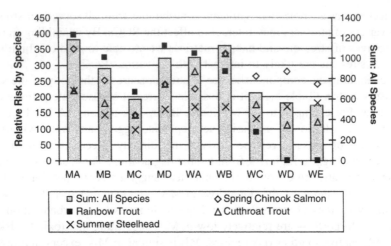

Figure 5.3 Cumulative and individual relative risk scores for salmonid habitat in each risk region.

that spring chinook salmon and rainbow trout have highest relative risk in MA, cutthroat trout have highest relative risk in WA, and summer steelhead have highest relative risk in WE.

By examining the RRM output on a species-by-species basis, stressor–habitat interactions are revealed. For example, Table 5.13 and Figure 5.3 show that the risk to spring chinook salmon is high throughout the project area, the highest risk being in the high-quality spawning area in risk region MA on the McKenzie River. The high risk ranks for spring chinook salmon in the Willamette River risk regions are due to the high value of the Willamette River in providing rearing habitat combined with increasing concentrations of sources of stress proceeding downstream in the project area.

Figure 5.4 shows further how the risk affects the habitat components for spring chinook salmon in the project area. Thus, it is apparent that risk in risk region MA is to all life stages whereas that in WB is to spawning and rearing habitat. By examining risk matrices for each species by source of stressor it is possible to predict stressor contributions to risks for individual species. The source-of-stressor matrix for chinook salmon (not shown), for example, is very similar to that of all species combined (Table 5.12), so similar stressor hypotheses to those discussed above for all species also apply to spring chinook salmon.

CONFIRMATION OF RISK RANKS

Risk ranks output from the RRM can be compared to generally accepted measures of environmental risk. These comparisons evaluate whether the pattern of predicted risk corresponds to the pattern of exposure and effect extant in the environment.

To test the model, ODEQ temperature, nutrients, BOD, total phosphorus, and total nitrates data were evaluated as indicators of water quality (EPA STORET 1995 to 1998

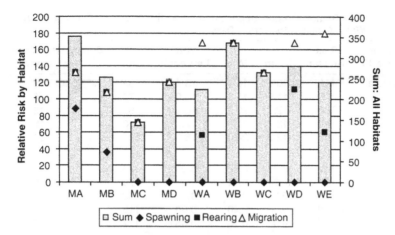

Figure 5.4 Relative risk scores for spring chinook salmon for each habitat type and cumulative risk scores for spring chinook in each risk region. Cumulative risk is the sum of the risk rankings for each habitat component.

data were retrieved for MA, and 1990 to 1998 data for all other sites). Oregon water quality criteria were used to evaluate divergence from acceptable conditions.

Divergence from optimal water quality acts as a stress on the aquatic community that when overlaid with receptors in habitats results in unacceptable risk of adverse effects. Thus, because poor water quality is an indication of stressors rather than risk, water quality was compared with RRM source ranks rather than risk ranks. Table 5.14 shows that water quality criteria were exceeded at all sites for all parameters except total nitrates, which exceeded threshold values only in risk regions WB and WE. The pattern of exceedances reflects generally decreasing water quality proceeding downstream. This pattern is consistent with the cumulative increase in anthropogenic landuse and is reflected in source rankings with an effects filter for water quality-related effects. Sources that release such stressors are landuse–agriculture, landuse–forestry, landuse–urban, water withdrawal, and NPDES.

One inconsistency between the source ranks and water quality data was encountered. Comparing values from Table 5.14 with ranks from Table 5.10 shows that for risk region WA, ranks appear inflated relative to instream measures of water quality. One possible explanation for this inconsistency is that the location of the water quality station in MA is in the upstream section of the risk region; thus, the influences of sources of stressors on water quality are not reflected in the available water quality data.

SENSITIVITY ANALYSIS

Sensitivity analysis provides a measure of uncertainty in the model results. Wiegers et al. (1998) designed a sensitivity analysis to quantify RRM model uncertainty. The RRM relies on the ability of the model to distinguish between random data and data specific to the risk components in the project area. When input is randomly chosen the model will not distinguish between risk regions, whereas input

Table 5.14   Risk Region Oregon Department of Environmental Quality (ODEQ)
             Water Quality Monitoring Data

Parameter (Benchmark)	Risk Region	Mean	Max	75th Percentile	n	Benchmark Exceedance Frequency (%)
Temperature	MA	10.4	18.4	13.1	34	3
(17.8°C)	MD	11.6	18.5	13.0	70	6
	WA	12.7	20.0	16.9	52	12
	WB	12.7	22.0	17.0	113	15
	WE	12.9	24.0	16.6	137	21
Total P	MA	0.04	0.07	0.04	30	7
(0.067 mg/L)	MD	0.04	0.14	0.05	70	10
	WA	0.04	0.14	0.02	52	10
	WB	0.06	0.40	0.06	113	20
	WE	0.07	0.18	0.08	137	50
Total nitrate	MA	0.035	0.10	0.04	31	0
(0.3 mg/L)	MD	0.058	0.18	0.09	70	0
	WA	0.067	0.25	0.12	52	0
	WB	0.127	0.35	0.17	113	1
	WE	0.294	2.20	0.40	137	25
BOD5	MA	0.69	1.4	0.9	30	3
(1.1 mg/L)	MD	0.99	2.5	1.3	70	39
	WA	1.04	2.3	1.3	52	40
	WB	1.14	3.1	1.4	113	43
	WE	1.08	2.5	1.8	137	40

that is risk related instead of random should drive the model to detect high-risk areas. To determine the sensitivity of the model, the results from randomly generated input were compared with those from input selected from a range of values representing uncertainty in the input values used in the project area.

To test sensitivity, the model was run 20 times with two different sets of input: all rank values were chosen randomly, and input was selected randomly from the range of possible uncertainty. For most sources of stressors data were subjectively analyzed for situations where the dispersion of data was small and small differences between data points caused them to be assigned to different ranks. In these cases, each risk region was assigned the next higher or lower rank, whichever was closer. For hatchery fish, where the ranks were assigned based on expert judgment, a range of ranks was assigned to each risk region based on a subjective assessment of the probability of hatchery fish interacting with native salmonids.

When the model was run with risk components randomly selected from the range uncertainty, the order of relative risk among the risk regions remained consistent with the original risk ranks. Out of 20 iterations, risk regions MA and WB were identified as having the highest risk and second highest risk rank 100% of the time. MC was identified as the lowest ranking risk region 80% of the time. Risk predictions generated from randomly assigned ranks identified MA and WA as the highest and second highest ranking risk regions, respectively, 15% of the time for each.

The sensitivity analysis shows that the results of the RRM are nonrandom predictions of risk, and that risk predictions correspond to real spatial variation in regional risk components in the upper Willamette Basin.

DISCUSSION

The Relative Risk Approach

This application of the RRM was in a large, open system with myriad disparate stressors acting on the ecology of the region. The overall pattern of risk was evident and risk matrices provided several hypotheses regarding the location of risks and their causes. Locations with high relative risks of multiple stressors interacting with valued resources were identified. Risk confirmation provided the first test of these hypotheses. Subsequent measurements of fish community structure, chemistry, and other elements of the project area will provide further tests.

There was considerable variation in the data available for characterizing different regional risk components. To ensure even treatment of the entire project area it was necessary to use data at a coarse resolution for most analyses. Data on physical habitat characteristics were generally not available even though some areas were well characterized. In the case of fish habitat and hatchery fish, poor data resolution required use of expert judgment rather than measured data to assign source of stressor and habitat ranks.

Lack of data on the physical habitat types and species-specific habitat use caused a divergence in the application of the RRM from the way that it was applied in Port Valdez. The Port Valdez assessment used common habitat types such as rocky shoreline, open water, shallow subtidal and mudflat habitats as a composite of unique receptors occupying each habitat. Ranks quantified the magnitude and quality of these habitats in each risk region. In applying the RRM to the upper Willamette Basin, critical lifestages of the ROCs were used as a measure of habitat quality. Habitat ranks were based on a subjective assessment of the ability of each risk region to support specific lifestages of each fish used as an assessment endpoint. This method enabled application of the RRM without the physical habitat data. In assigning habitat values to the upper Willamette Basin, the ROCs were highly mobile with their habitat use varying considerably throughout their life cycle. Thus, predicting the risk of population level declines depends on assessing risk to various life stages in multiple habitats. Because quantification of habitat values requires detailed physical data which were not available, subjective data were deemed most appropriate for the scale and resolution of this RRM application.

In the upper Willamette Basin the risk regions were arranged along the course of the river where materials flow downstream. In this case each risk region strongly affects subsequent risk regions. By assigning risk regions based on the watershed contributing to distinct stretches of the river, the risk components affecting unique provinces within the project area were logically aggregated and watershed approach provided a way to account for the flow of materials downstream.

Risk Confirmation

By comparing standard measures of risk with the risk scores appropriately filtered for effects, the magnitude of stressors represented by the source of stressors ranks in each risk region was confirmed. As with the Port Valdez assessment, lack of relevant data taken over the entire project in a systematic fashion limited the confirmation of the risk ranks to just a few of the risk regions.

Because of the type of data available, the risk confirmation is only partial. Though the water quality data represented the best data available for risk confirmation, exceedence of benchmark values for these parameters does not represent the full suite of stressors acting on the receptors. In the natural environment, additive or synergistic effects with other chemical, physical, or biological stressors occur, compounding effects predicted based on benchmark values. The RRM relies on the assumption that the relative magnitude of any interactions between stressors increases with the relative magnitude of the sources producing those stressors. Accounting for only a portion of the stressors predicted to be present through risk confirmation provides only a partial picture of the total risk. Thus, these standard measures of risk provide a rough baseline for the risk scores. Ongoing sampling of the fish community structure and other elements of the aquatic community will more completely characterize the risk and provide experimental evidence to further test risk predictions.

Uncertainty

Two types of uncertainty arise in risk assessment: systematic errors and random errors (Seiler and Alvarez 1998). These can further be distinguished into uncertainty in problem formulation, model uncertainty, and parameter uncertainty (EPA 1998b). In the RRM, error in the problem formulation results when sources of stressors in the project area were not identified or important aspects of the ecology were not developed. This uncertainty is ameliorated somewhat in that NCASI and ODFW personnel were consulted throughout the problem formulation, thus some peer review occurred. Other stakeholders were not directly consulted; instead, published accounts of stakeholder opinions such as the Willamette Valley Livability Forum (WVLF 1999) and the WRBS (OWRD 1999a) were relied on to represent their viewpoints in establishing assessment endpoints, receptors of concern, and the conceptual models. Therefore, results of the RRM may not address some potential concerns for the project area.

Model uncertainty includes errors promulgated through faulty assumptions in the individual analyses and in the reassignment of the original data to ranked data. In a risk assessment of this scale it was necessary to use course scale data collected for other purposes. Thus, important temporal and spatial variation in the environment may have been overlooked.

Risk predicted by the RRM is relative. Thus, the relative risk within a given risk region may be the highest in the project area, but still below a threshold where it is unacceptable. In order to provide a baseline for the magnitude of risk represented by the risk predictions, confirmatory analysis using data not included in the original analyses should be conducted. This would enable comparison to standardized measures of risk. As a partial confirmation, water quality data collected by ODEQ generally concurred with the source of stressors ranks input into the RRM.

UTILITY

The RRM provides a means of combining multiple stressors and receptors in a regional risk context without using probability estimates that rely on an assumption of the underlying relationship linking disparate stressors and receptors for multiple assessment endpoints. By using a relative ranking approach instead of probability estimates, the RRM is capable of discerning where the potential for effects is highest and identifying multiple stressors and receptors that may interact to cause adverse effects. The RRM suggests the types of stressors that may interact and the locations where they are most likely to adversely affect ROCs. Thus, it provides a useful tool for identifying adverse effects and the locations where they may be occurring.

Through the use of more detailed data and site-specific process models, uncertainty could be reduced and the RRM could be used for a higher-tiered risk assessment. In the upper Willamette Basin the greatest reductions in uncertainty could be gained through detailed information on physical habitat and its use by the salmonids. Improvements in uncertainty regarding each of the stressors acting on the ecology of the region could also be reduced through detailed data regarding the processes releasing the stressors. For example, uncertainty regarding toxicity-associated stressors could be reduced through application of detailed fate and transport models.

REFERENCES

ACOE (U.S. Army Corps of Engineers and Oregon State University). 1999a. A spatial representation of the U.S. Army Corps of Engineers' database regarding revetments along the Willamette River, Oregon. Unpublished.

ACOE (U.S. Army Corps of Engineers). 1999b. Revetment data for the Willamette and McKenzie Rivers, Oregon. Unpublished.

ACOE (U.S. Army Corps of Engineers and Oregon State University Department of Fish & Wildlife). 1998. A spatial representation of the Willamette River Inundation, Flood of 1996.

Altman, B., Henson, C.M., and Waite, I.R. 1997. Summary of Information on Aquatic Biota and Their Habitats in the Willamette Basin, Oregon, through 1995. U.S. Geological Survey Water-Resources Investigations Report 97-4023.

Bonn, B. 1998. Dioxins and furans in bed sediment and fish tissue of the Willamette Basin, Oregon, 1992–95. U.S. Geological Survey, Water-Resources Investigations Report 97-4082-D.

EPA (United States Environmental Protection Agency). 1999. Envirofacts warehouse, toxic release inventory (input specific facility ID). http://www.epa.gov/enviro/html/tris/tris_query.html (accessed June 15, 1999).

EPA (United States Environmental Protection Agency). 1998a. Better Assessment Science Integrating Point and Nonpoint Sources: BASINS Version 2.0. United States Environmental Protection Agency, Office of Water. EPA-823-B-98-006.

116   REGIONAL SCALE ECOLOGICAL RISK ASSESSMENT

EPA (United States Environmental Protection Agency). 1998b. Guidelines for Ecological Risk
    Assessment. Risk Assessment Forum, Washington, D.C. EPA/630/R95/OOZF. April.
Groop, R.E. 1980. JENKS: An Optimal Data Classification Program for Choropleth Mapping.
    Technical Report 3. Department of Geography, Michigan State University, March,
    1980, in *Cartography: Thematic Map Design*, Dent, B.D., Ed., 1990. Wm. C. Brown,
    Dubuque, IA.
Kammerer, J.C. 1990. Largest Rivers in the United States: U.S. Geological Survey Open–file
    Report 87-242.
Laenen, A. and Risley, J.C. 1997. Precipitation-Runoff and Streamflow-Routing Models for
    the Willamette River Basin, Oregon. U.S. Geological Survey Water-Resources Inves-
    tigations Report 95-4284. Prepared in cooperation with the Oregon Department of
    Environmental Quality, Portland, OR.
Landis, W. G. and J. A. Wiegers. 1997. Design considerations and a suggested approach for
    regional and comparative ecological risk assessment, *Hum. Ecol. Risk Assess.*, 3,
    287–297.
ODEQ (Oregon Department of Environmental Quality). 1998. Oregon's Final 1998 Water
    Quality Limited Streams – 303(d) List. http://www.waterquality.deq.state.or.us/wq
    (accessed November 19, 1998).
ODFW (Oregon Department of Fish and Wildlife). 1998. Spring Chinook Chapters Willamette
    Basin Fish Management Plan. Oregon Department of Fish and Wildlife, Portland, OR.
ODFW (Oregon Department of Fish and Wildlife). 1991. Willamette Basin Fish Management
    Plan, Willamette River Subbasin. Oregon Department of Fish and Wildlife, Portland,
    OR.
ODFW (Oregon Department of Fish and Wildlife). 1990. Willamette Basin Salmon and
    Steelhead Production Plan (includes separate reports for Willamette River Basin and
    ten subbasins): Portland, Columbia Basin system planning, Oregon Department of
    Fish and Wildlife, Portland, OR.
ODFW (Oregon Department of Fish and Wildlife). 1988. Willamette Basin Fish Management
    Plan: Oregon Department of Fish and Wildlife, Portland, OR.
OWRD (Oregon Water Resources Department). 1999a. Willamette Basin Reservoir Study.
    http://www.wrd.state.or.uspublication/reports/reports.html (accessed December
    21,1998).
OWRD (Oregon Water Resources Department). 1999b. Water Availability Resources System.
    unpublished. telnet://wars.wrd.state.or.us (accessed June 18, 1999).
Seiler, F.A. and Alvarez, J.L. 1998. Letter to the editor: on the use of the term uncertainty,
    Human Ecol. Risk Assess., 4, 1041–1043.
Suter, G.W. 1990. Endpoints for regional ecological risk assessments, *Environ. Manage.*, 14,
    9–23.
USGS. 1999. http://www.oregon.wr.usgs.gov.html (accessed June 8, 1999).
Wade, M. 1999a. Mark Wade ODFW fisheries biologist. Personal communication with Matt
    Luxon. Phone conversation and e-mail exchange regarding habitat ranking for salmo-
    nids in the project area.
Wade, M. 1999b. Mark Wade ODFW fisheries biologist. Personal communication with Matt
    Luxon. Willamette basin hatchery release records.
Wentz, D.A., Bonn B.A., Carpenter, K.D., Hinkle, S.R., Janet, M.L., Rinella, F.A., Uhrich,
    M.A., Waite, I.R., Laenen, A., and Bencala, K.E. 1998. Water Quality in the Wil-
    lamette Basin, Oregon, 1991–95. U.S. Geological Survey Circular 1161.
Wiegers, J.K., Feder, H.M., Mortenson, L.S., Shaw, D.G., Wilson, V.J., and Landis, W.G.
    1998. A regional multiple-stressor rank-based ecological risk assessment for the fjord
    of Port Valdez, Alaska, *Hum. Ecol. Risk Assess.*, 4, 1125–1173.

Willamette Basin Task Force. 1969. Willamette Basin comprehensive study – Water and related land resources, Appendix D – Fish and Wildlife: Willamette Basin Task Force, Pacific Northwest River Basins Committee Report.

WVLF (Willamette Valley Livability Forum). 1999. http://www.econ.state.or.us/wvlf/visfinl. html.water (accessed August 17, 1999). Revised version available at http://www.wvlf. org/facts.html.

CHAPTER 6

Codorus Creek Watershed: A Regional Ecological Risk Assessment with Field Confirmation of the Risk Patterns

Angela M. Obery, Jill F. Thomas, and Wayne G. Landis

CONTENTS

INTRODUCTION

The risk assessment for Codorus Creek was the second regional-scale risk assessment using the relative risk model (RRM) to be published (Obery and Landis 2002). The Codorus Creek watershed (CCW) in Pennsylvania is an excellent example of the challenge of performing risk assessments at this scale and with multiple types of stressors. Located within the watershed are a paper mill, a growing urban area, agriculture, recreational fishing, and the water source for the City of York. There are multiple groups of interested parties including a watershed association, state regulatory agencies, the City of York and other towns, sports fishermen, and local citizens.

To this scenario we applied environmental risk assessment as a data interpretation and decision-making tool. Because of the size of the area of interest we conducted a regional risk assessment using the RRM in order to incorporate these multiple sources, stressors, and endpoints measures of risk (Obery and Landis 2002). The overall patterns of the risk were then confirmed by field research that examined both the fish and macrobenthic assemblages. Finally, a set of alternative management schemes was evaluated and the changes to the risk pattern analyzed.

In this chapter we introduce the RRM ecological risk assessment (EcoRA) and summarize results of the field studies as presented in Obery and Landis (2002). The remainder of the chapter discusses the confirmation of the risk patterns from the multivariate analysis of the field data not used in the initial risk assessment. The use of the RRM in evaluating management strategies in altering the risk within the CCW is presented in Chapter 7.

Regional Risk Assessment and the Relative Risk Model (RRM)

The RRM was developed in order to integrate the impacts due to a variety of stressors at a regional scale (Landis and Wiegers 1997; Wiegers et al. 1998; Chapters 1 and 2 of this volume). The RRM has been used successfully at a variety of sites including Valdez, Alaska; Mountain River, Tasmania (Walker et al. 2001); and the PETAR reserve in Brazil (Moraes et al. 2002). The basic premise of the method is the innate consideration of (1) the interactions between sources of stressors, habitats, and endpoints, (2) where these interactions occur in a geographical context, and (3) the use of ranks to describe the risk that results from these spatial interactions. Introductions to the RRM have been published (Landis and Wiegers 1997; Landis and Yu 2004), and the calculations and means of presenting uncertainty detailed (Wiegers et al. 1998; Obery and Landis 2002).

In a regional risk assessment conceptual model there has been a source that releases a set of stressors; the stressors are transmitted to a specific habitat that is the home to a group of receptors. Exposure to these receptors results in a series of predicted impacts. It is understood that there are multiple sources of various stressors, that a variety of habitats may exist, and that multiple responses may occur. Central to this approach is that each source, habitat, and impact has a location in the study area and an associated map coordinate.

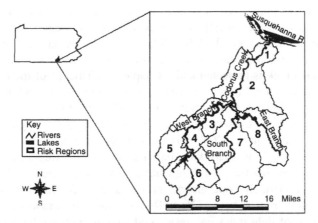

Figure 6.1 The Codorus Creek Watershed and the risk regions. (From Obery, A.M. and Landis, W.G., *Hum. Ecol. Risk Assess.*, 8, 405–428, 2002. With permission of Amherst Scientific Publishers.)

Part of the RRM involves mapping the locations of sources, stressors, habitats, and impacts. Without spatial overlap there is no causality and no likelihood of an observed impact. Stressors can be differentiated by where they occur. Our regional approach incorporates a system of numerical ranks and weighting factors to address the difficulties encountered when attempting to combine different kinds of risks. Ranks and weighting factors are unitless measures that operate under different limitations than measurements with units (e.g., mg/L, individuals/cm^2). We link these ranks to specific locations within a landscape, providing a map with the relative risks ranked from low to high.

PART I: THE CODORUS CREEK WATERSHED AND THE REGIONAL RISK ASSESSMENT

The Codorus Creek Watershed

The study boundary is the entire CCW, located in south central Pennsylvania. The CCW drains an area of 719 km^2 (278 mi^2) in York County (Figure 6.1). The creek flows 77 km (48 mi) in a northeasterly direction from the longest tributary to the discharge into the Susquehanna River. The entire watershed contains 596 km of creek bed, and perennial streams range from less than a meter wide to approximately 36 mi wide. The watershed extends from the Codorus Creek headwaters with three main tributaries, referred to as the East Branch, South Branch, and West Branch, to its confluence with the Susquehanna River near Harrisburg, PA. As a subbasin of the Lower Susquehanna River and a tributary to the Chesapeake Bay, the drainage area is highly developed in terms of population, industrial centers, and productive agricultural area and has undergone a high level of scrutiny. The watershed contains urban and rural communities including York, Spring Grove, and Hanover.

Codorus Creek is designated as a *priority* water body due to the presence of a public water supply in the watershed, documentation of toxicity related to fish and aquatic life in the watershed (USGS 1999), and the presence of major National Pollutant Discharge Elimination System permits in the watershed (SRBC 1991). Water use of the creek is protected under Chapter 93, Title 25 of the Pennsylvania code for statewide general use, trout fishery, warmwater fishery, coldwater fishery, and high-quality coldwater fishery water use (PADEP 1998).

For the assessment, stressors were organized into eight risk regions according to their spatial position in the CCW (Figure 6.1). Risk regions were determined by grouping subwatersheds by areas with similar landuse. Risk Region 1 is composed of watersheds that lie between the Susquehanna River and the city of York in what is considered a moderately undeveloped rural area. Risk Region 2 is composed of subwatersheds that contain most of York. Risk Region 3 is composed of subwatersheds that consist of light industrial, residential, and agricultural landuse just south of Indian Rock Dam and 0.8 mi north of the Highway 116 bridge and includes the industrial waste discharges from a pulp and paper mill. Risk Region 4 is composed of subwatersheds south of Region 3 and includes the Menges Mill community at the southwestern boundary and the Kraft Mill community at the southern boundary. Risk Region 5 is composed of the Oil Creek subwatershed and consists of the Glooming Grove community, rural residences, and agriculture. Risk Region 6 is composed of subwatersheds that contain Lake Marburg and West Branch and consists of residential and agricultural landuse. Risk Region 7 is composed of subwatersheds that drain into South Branch, and Region 8 is represented by the subwatersheds that drain into East Branch. Risk Regions 7 and 8 contain primarily rural residential and agricultural landuse, with Region 8 containing the primary drinking water supply for York County.

The ecological assessment endpoints were selected after a Codorus Creek Watershed Association meeting that included representatives from various stakeholder groups such as the Pennsylvania Department of Environmental Protection (PADEP), local industries, Trout Unlimited, and local citizens. The assessment endpoints were:

1. Protective water quality for aquatic ecological receptors and humans during contact or consumption
2. Adequate water supply for drinking and waste discharge
3. Self-sustaining native and nonnative fish populations in the watershed
4. Adequate food availability for aquatic species
5. Available recreational land and water resources
6. Adequate stormwater control and treatment

Conceptual Site Model

An ecological conceptual site model (CSM) was developed to represent the general relationships between the stressors and the assessment endpoints that constitute the primary exposure pathways assessed in the CCW regional EcoRA (Figure 6.2). The CSM was developed from information about the identified sources of stress (i.e., stressors), potential exposure pathways, and predicted effects on endpoints. As evident in the CSM, multiple stressors and exposure pathways are present.

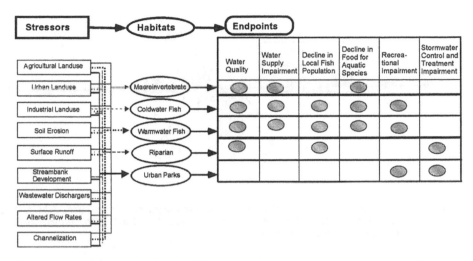

Figure 6.2   Conceptual site model (CSM) for the interaction between stressors, habitats, and assessment endpoints. This is the conceptual model used for the original risk assessment and then for management scenarios. (From Obery, A.M. and Landis, W.G., *Hum. Ecol. Risk Assess.*, 8, 405–428, 2002. With permission of Amherst Scientific Publishers.)

Relative EcoRA compares stressors and habitats in risk regions and determines if the chance of an impact is greater in one risk region than another. Ranks, also referred to as comparative risk estimates, are unitless values that show the locations with the greatest probability of impacts to valued endpoints. Relative risk estimates are based on the following assumptions (Landis and Wiegers 1997; Wiegers et al. 1998):

1. The greater the relative distribution of a stressor to the risk region area, the greater the potential for exposure to habitats in that risk region.
2. Stressors are limited to those with the greatest potential for adverse impacts.
3. For an assessment endpoint to be adversely impacted, there must be a complete exposure pathway from the stressor to the habitat.
4. Multiple stressors that impact assessment endpoints are additive in their relative ranks. This assumption was made out of convenience and lack of knowledge and literature.
5. Surrogate data applied in place of actual stressor measurements and habitat-monitoring data are representative of site conditions.

Risk characterization was used to rank complete exposure pathways established in the CSM to the endpoint selected for each risk region. Relative ecological ranks were summarized by the sum of relative ranks per stressor, sum of relative ranks per habitat, sum of relative risks per endpoint, and relative risk per risk area.

Overall Risk Ranks

Figure 6.3 provides a summary of overall risk ranks for regions in the CCW, and the scores are found in Table 6.1. Referring to the total endpoint rank, Table

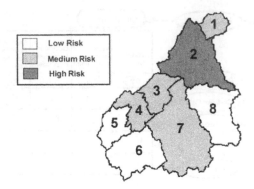

Figure 6.3   Distribution of risk within the Codorus Creek watershed. (From Obery, A.M. and
Landis, W.G., *Hum. Ecol. Risk Assess.*, 8, 405–428, 2002. With permission of
Amherst Scientific Publishers.)

Table 6.1   Risk Scoring for Endpoints and Risk Regions from the CCW EcoRA.

Risk Region	Sampling Site	Endpoint: Decline in Local Fish Population	Endpoint: Decline in Food Availability for Aquatic Species	Total Risk Region Rank	Final Risk Classification
1	Furnace Bridge	336	396	2004	Medium
2	Arsenal Bridge	292	480	2508	High
2	Indian Rock Dam	292	—	2508	High
3	Graybill Bridge	328	448	2048	Medium
3	Martin Bridge	328	448	2048	Medium
3	USGS Gauging Station	328	448	2048	Medium
4	Menges Mill	348	468	2136	Medium
5	None	—	—	—	Low
6	Park Road	264	324	1676	Low
7	None	—	—	—	Medium
8	None	—	—	—	Low

Source: Obery, A.M. and Landis, W.G., *Hum. Ecol. Risk Assess.*, 8, 405–428, 2002. With
permission.

6.1 illustrates that water quality impairment is the assessment endpoint at greatest
risk for the entire watershed, with the greatest impact occurring in Region 2. Region
2 demonstrates the largest overall risk and Region 8 demonstrates the smallest overall
risk. Jenk's optimization clustered the total of risk region ranks as low risk (Regions
5, 6, and 8), medium risk (Regions 1, 3, 4, and 7), and high risk (Region 2). A
detailed analysis is supplied in Obery and Landis (2002).

PART II: VERIFICATION OF RELATIVE RISK CLASSIFICATIONS

Verification of the pattern of risk scores for the fish population and macroinvertebrate population endpoints was achieved. First, fish and macroinvertebrates were
collected independently and an assemblage dataset was constructed. Second, three

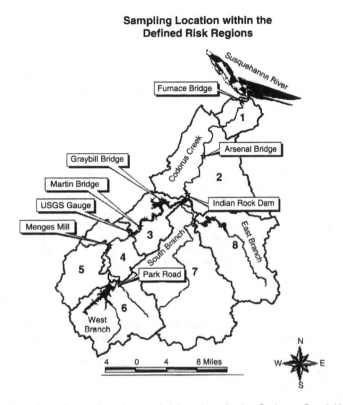

Figure 6.4   Location of sampling sites and risk regions in the Codorus Creek Watershed in south central Pennsylvania.

multivariate statistical methods (principal components analysis, hierarchical cluster-ing, and discriminant analysis) were employed to compare the resulting patterns to the patterns of risk generated by the EcoRA. The patterns between the risk assess-ment scores corresponded to the observed upstream-to-downstream gradients and in the outliers (Thomas 2001) for both datasets.

Biological Datasets

We made use of two biological datasets in this study. Western Washington University (WWU), as part of the ongoing long-term receiving waters study (LTRWS) being performed by the National Council for Air and Stream Improvement (NCASI) (NCASI 2002; 2003), generated the fish community dataset. The macro-invertebrate community dataset was generated by NCASI also as a part of the LTRWS (NCASI 2002; 2003).

Teams made up from WWU and NCASI personnel gathered the fish community data. They sampled on a quarterly basis from six sites along the West Branch of Codorus Creek and two sites on the main stem of Codorus Creek downstream of the confluence of the three tributaries (Figure 6.4, Table 6.2). The subset data used in this analysis consisted of six sampling dates covering an 18-month period from

Table 6.2   The CCW EcoRA Risk Region and Sampling Site Descriptions

Risk Region	Area Description	Landuse	Fisheries Type	Biological Sampling Sites with Distance from Pulp Mill outfall (– upstream, + downstream)
1	Subwatersheds that lie between the Susquehanna River and the city of York	Moderately undeveloped rural area	Warm water	Furnace Bridge
2	Subwatersheds that contain most of the city of York	Highly urban and industrialized area	Warm water	Arsenal Bridge Indian Rock Dam
3	Subwatersheds bounded by Indian Rock Dam at the S. and P.H. Glatfelter Pulp and Paper Mill to the N.	Light industrial (pulp and paper mill effluent), residential and agricultural area	Warm water	Graybill Bridge (+10 river km) Martin Bridge (+2.2 river km) USGS Gauging Station (−1.0 river km)
4	Subwatersheds of Spring Grove bounded by P.H. Glatfelter to the N., Menges Mill to the S.W. and Kraft Mill to the S.	Rural residential, agricultural area	Warm water N. of Menges Mill, cold water S. of Menges Mill	Menges Mill (−5.3 river km)
5	Subwatershed of Oil Creek	Glooming Grove, rural residential and agricultural area	Not identified	No sampling sites
6	Subwatersheds containing Lake Marburg and West Branch headwaters	Rural residential and agricultural area	Cold water at the outlet of Marburg Dam, Trout Fisheries S. to Headwaters	Park Road
7	Subwatersheds of the South Branch	Rural residential and agricultural area	Not identified	No sampling sites
8	Subwatersheds of the East Branch	Rural residential and agricultural area, primary drinking water supply for York County	Not identified	No sampling sites

September 1998 through March 2000, inclusive. Electrofishing by a three-person team was used to sample the fish, with one person electroshocking and two people netting fish. Each site had three runs of approximately 600 seconds each for a total sampling time of 1800 seconds. The team identified the collected fish to family and to species where possible, took weights and measurements, and the fish were then released. All fish not identified by the team on site were frozen and transported to WWU for later identification and measurement.

Macroinvertebrates were collected from five sites along the West Branch of Codorus Creek and two sites along the main stem. The subset of data used in this analysis consisted of five sampling dates covering a 15-month period from September of 1998 to November of 1999, inclusive. A three-person team using Surber, Kicknet, or Hess equipment and making three to five repetitions sampled the macroinvertebrates. The collected macroinvertebrates were preserved in ethanol, formalin, or an ethanol–formalin mixture and transported to an outside consultant for taxonomic identification to order, family, and genus. Additional information was derived for richness, tolerance, feeding group, and community diversity measurements.

Fish and macroinvertebrate samples were collected within 3 weeks of each other, with macroinvertebrate sampling occurring first over a 2-day period followed by fish sampling over 2 days. All sampling was done in a downstream-to-upstream direction.

Data Analysis

All data analysis was performed using the SPSS Base 9.0 data analysis program (Chicago, IL). The raw data for fish were standardized to three passes at 600 seconds each and then sorted to number of individuals per species per site for each sampling date. When fish could not be identified to species, family designations were used. Macroinvertebrate raw data were sorted to number of individuals per genus per site for each sampling date. When identity to genus was not available, identification to family or order was used. Descriptive statistics were run on the total sample for each group by site, by date, and by taxa. Fish and macroinvertebrate data were determined to have nonnormal distributions using the Shapiro–Wilk's test, and nonequality of variance was determined using the Levene test. A spread-vs.-level plot was used to determine the best possible method of transformation for each group. We used principal components analysis (PCA) on the raw and transformed fish and macroinvertebrate data to identify trends for comparison to the CCW EcoRA. We used hierarchical clustering on the raw data and discriminant analysis on the transformed data to confirm the patterns observed in our PCA results.

PCA is particularly useful for exploration of linear environmental gradients (Sparks et al. 1999). While PCA does not require normal data, nonnormal data may distort results. In order to evaluate any possible distortion we square root transformed the fish data and log transformed the macroinvertebrate data and reran the PCA. Both analyses gave similar site separation patterns as the nontransformed data. Based on this we believe that no significant distortions occurred when using the nontransformed data. PCA assumes a linear relationship; therefore, we first determined that all fish and macroinvertebrates used in the PCA analysis were significantly correlated ($\alpha = 0.05$) to site by nonparametric Spearman's ρ and/or Kendall's τ-b methods. PCA was run without rotation for the fish and macroinvertebrate datasets individually and when combined. We maximized for clearest separation of sites with the greatest explanation of variance, eliminating variables that had low correlations and low loading. Trends along sites were then compared to trends in the CCW EcoRA rankings for decline in fish populations and food for fish populations.

We performed hierarchical clustering on the fish and macroinvertebrate taxa that had resulted in the best separation of sites by the PCA analysis. We also ran the analysis on all the fish and macroinvertebrate taxa that were significantly ($\alpha = 0.05$) nonparametrically correlated with the site. For the hierarchical clustering we used three different measures of distance (euclidean, squared euclidean, and cosine) with seven different methods of clustering (average within groups, average between groups, nearest neighbor, furthest neighbor, centroid, median, and Wards). We ran all possible distance-clustering combinations on the taxa counts and on a binary (presence/absence) version of the dataset.

To evaluate the predictive nature of the separations we ran discriminant analysis using Wilk's lambda stepwise method on the PCA-selected fish taxa and the PCA-selected macroinvertebrate taxa. We square root transformed the fish data and used those cases (11 of 12) that met the assumptions of within-site normal distribution and heterogeneity of variance. We log transformed the macroinvertebrate data and used those cases (10 of 12) that met the assumptions. We ran leave-one-out analysis and a training set analysis in which we split the dataset into two groups, using the first group as a training set to test the classification of the second group. The first group for the fish consisted of the first four sampling dates; the first group for the macroinvertebrates consisted of the first three sampling dates. For both datasets the unselected second group consisted of the last three sampling dates. We used predetermined classification groupings based on our PCA and clustering results and the CCW EcoRA risk scores and risk regions.

Index of Biotic Integrity (IBI)

We calculated indices of biotic integrity for fish and used a provided IBI for the macroinvertebrates for comparison to our multivariate analyses and the RRM EcoRA results. We modified the Warmwater Streams of Wisconsin fish IBI (Lyons 1992) per an earlier Codorus Creek biological assessment (Snyder et al. 1996) using 10 of 12 metrics for the warm water sampling sites (Furnace Bridge, Arsenal Bridge, Indian Rock Dam, Graybill Bridge, Martin Bridge, and USGS) and for the sites at (Menges Mill) and above (Park Road) the cold water reach which may have influences of both warmwater and coldwater aspects. The modification from Snyder et al. (1996) replaced the number of sucker species with the number of minnow species. We did not collect information on fish condition, so the final metric of proportion of diseased or anomalous fish could not be analyzed. We also excluded the fish density metric due to low overall catch-per-unit rates for our study area. To ascertain the significance of omitting these two metrics, a sensitivity analysis was performed for both metrics. The sensitivity analysis consisted of calculating the IBI scores using the highest possible value for the missing metric and repeating the process using the lowest possible value. The resulting range and patterns of distribution were not substantially different from the calculations made without the missing metrics. Based on the sensitivity analysis we do not believe the between-site relationships in the warmwater IBI were significantly altered by these modifications. Sites with a score of 20 or less were rated as poor, scores of 22 to 32 were rated fair, and scores of 33 or greater were rated good. The scores were plotted and trends were compared to the CCW EcoRA risk ranking trends and the trends generated by the multivariate statistical analyses.

We used a modified coldwater fish IBI (Lyons et al. 1996) with four metrics. The modification consisted of eliminating the metric that measured percent of salmonids as brook trout, as they do not normally occur in this habitat. The coldwater IBI was run on four sites that are potentially impacted by coldwater: Park Road (upstream from the hypolimnetic discharge from Lake Marburg Dam, temperature range during sampling of 6.5 to 20.0°C), Menges Mill (at the juncture of the warmwater and coldwater stretches, temperature range during sampling of 8.0 to 17.0°C), USGS Gauging Station (downstream from Menges Mill, temperature range during sampling of 9.5 to 24.4°C), and Indian Rock Dam (downstream from the confluence of the three branches, including a coldwater stretch of the East Branch, temperature range during sampling of 7.7 to 23.5°C). We evaluated scores of 8 to 16 as poor, 24 to 40 as fair, 48 to 64 as good, and 72 to 80 as excellent. The scores were evaluated for any trends and compared to the CCW EcoRA risk ranking trends and the trends generated by the multivariate statistical analyses.

A macroinvertebrate Hilsenhoff biotic index (HBI) was calculated for all macroinvertebrate sites by an outside source (NCASI 2002) and used in our trend analysis. We used two separate evaluation criteria for the HBI results. The lower evaluation criteria (Matthews et al. 1998) rated sites with scores less than 1.75 as clean and sites with scores greater than 3.75 as polluted. The higher evaluation criteria (Lyons et al. 1996) rated scores less than 5.01 as approximating subecoregional reference value, scores of 5.01 to 6.26 as deviating somewhat from reference value, and scores greater than 6.26 as deviating strongly from reference value. The trends were then compared to the risk rankings for the endpoint of decline in food availability for aquatic species.

Uncertainties

Each method introduced a level of uncertainty to the outcome. The sampling site selection introduced uncertainty in the choice and location of the sites. The absence of sampling sites in three of the eight subregions left us unable to evaluate the risks for these three regions. Additionally, sampling sites were selected as representative of fish habitat and so were not necessarily typical or representative of all types of habitat in the watershed. This may have introduced a bias in the data collected.

Possible areas of uncertainty introduced in the fish sampling included variability in sampling times and runs, weather (i.e., increasing turbidity and flow rates), data gaps due to equipment failure, methodology (the unequal effect on different fish species by electroshocking), and personnel changes in the three-person sampling team. Possible areas of uncertainty introduced in the macroinvertebrate sampling were variability in sampling equipment, preservation techniques, replications, and methodology (possible nonrepresentative sampling).

Areas of uncertainty introduced during the data analysis included the initial processing and the analysis. The step standardizing the fish data to sample time

Table 6.3   Biological Variables of Importance for Separating Sites along Codorus Creek

Fish Species	Macroinvertebrate Taxa
Banded Darter (*Etheostoma zonale*)	*Coleoptera Elmidae Dubiraphia*
Blacknose Dace (*Rhinichthys atratulus*)	*Coleoptera Elmidae Stenelmis*
Brown Trout (*Salmo trutta trutta*)	*Coleoptera Psephenidae Psephenus*
Creek Chub (*Semotilus atromaculatus*)	*Diptera Chironomidae Dicrotendipes*
Fathead Minnow (*Pimephales promelas*)	*Diptera Chironomidae Microtendipes*
Greenside Darter (*Etheostoma blennioides*)	*Diptera Chironomidae Parametriocnemus*
Longnose Dace (*Rhinichthys cataractae*)	*Diptera Chironomidae Paratanytarsus*
Margined Madtom (*Notorus insignis*)	*Diptera Chironomidae Stempellinella*
Pumpkinseed (*Lepomis gibbosus*)	*Ephemeroptera Ephemerellidae Serratella*
Rock Bass (*Ambloplites rupestris*)	*Ephemeroptera Tricorythidae Tricorythodes*
Smallmouth Bass (*Micropterus dolomieui*)	*Trichoptera Hydroptilidae Hydroptila*
White Sucker (*Catostomus commersoni*)	*Trichoptera Psychomyiidae Psychomyia*

assumed a linear relationship between time and number of fish caught. This may have resulted in lower or higher numbers than would have actually occurred. Additionally, standardizing the fish data to 1800 seconds per sample had an unequal effect depending on the numbers of a given species present, with species with large numbers of fish being affected more than species with only one or two fish present per site.

Uncertainty was introduced in the index analysis by using fish IBIs developed for a midwestern region. Even with modifications it may not have given an accurate measurement. Another area of introduced uncertainty for the warmwater IBI was using the scoring criteria from the earlier Codorus Creek biological assessment (Snyder et al. 1996). This was necessary, as we did not have a reference dataset or a reference site for this study. Additionally, leaving out the fish condition metric and the catch-per-unit-effort metric may have altered the between-site relationships. The coldwater IBI had uncertainty introduced by the reduction of the five metrics to four, making each metric much more powerful. This would have the probable result of diminishing the ability to distinguish small differences between sites.

Summary of Verification Results

Fish Population Analysis

The original dataset for the fish assemblages consisted of 46 categories identified to species or family. In order to evaluate trends we used a subset of 19 fish species that we found to be significantly nonparametrically correlated with site.

Our PCA on untransformed data identified 12 fish species (Table 6.3) that allowed us to separate seven of the eight sampling sites using the first three components, and explained 72% of the variation. Plotting the first and third components, accounting for over 48% of the variation, allows for clear separation of the two most upstream sites and the two most downstream sites (Figure 6.5a). When the area between −1 and 1 is expanded for each component, three of the four inner sites are visibly separated, with the fourth site overlapping with the three immediate downstream sites (Figure 6.5b). Plotting against the second component shows similar

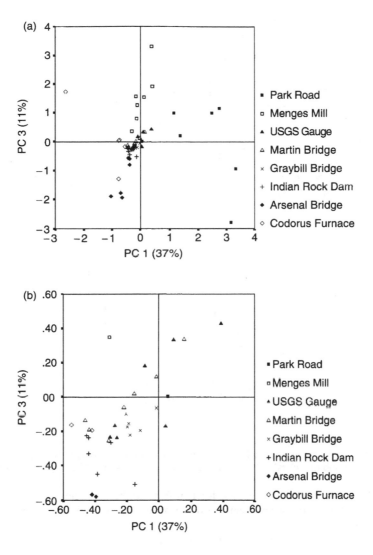

Figure 6.5 Scatter plot of the first and third factor scores from PCA analysis on the fish community data: (a) plot showing all the data points, (b) plot showing an expanded region from −.60 to 0.60.

separation (not shown). When the mean factor scores for each component are plotted by site, a clear upstream-to-downstream gradient is evident, with some deviation from the trend by the furthest upstream and downstream sites (Figure 6.6a, b). These gradients show that the inner sites have less within-site and between-site variation than the outer sites.

The clustering analysis had variations in clustering outcome between the different methods used, but similarities appearing across the methods indicated three strong trends, all of which supported the PCA:

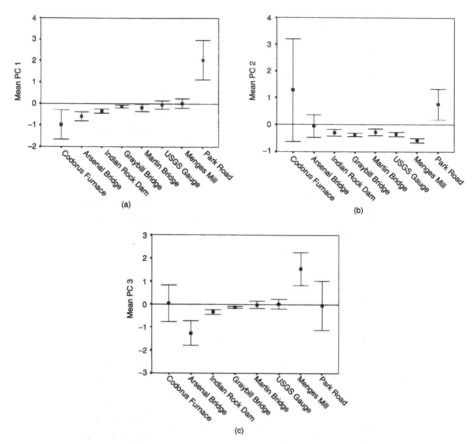

Figure 6.6 Mean factor scores from PCA on the fish dataset plotted for each site (+/– 2 standard errors): (a) factor score PC 1, (b) factor score PC 2, (c) factor score PC 3.

1. The most upstream site is very different from all the other sites.
2. The two outermost upstream and downstream sites are different from the four inner sites.
3. The four inner sites are very similar to each other.

Discriminant analysis was run on square root transformed data from 11 of the 12 fish species from the PCA analysis. Fathead minnow was excluded from this phase of the analysis because it could not be transformed to give a normal within-site distribution and heterogeneity of variance. The program was asked to classify the sampling units based on five different sets of classifications; the first and second were suggested by the PCA and clustering analysis, the third and fourth were based on the CCW EcoRA, and the fifth classification set was based on the eight sampling sites. The classifications were:

1. Three groups consisting of the two upstream sites, the four middle sites, and the two downstream sites
2. Four groups the same as the first, but splitting the two upstream sites into separate groups

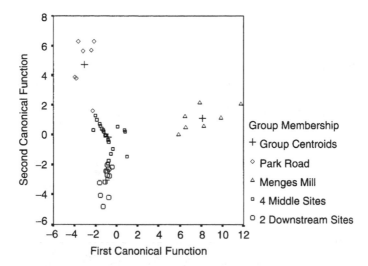

Figure 6.7   First two canonical functions plotted for 11 fish using the leave-one-out method for the second classification system of four regions: two separate upstream sites, combined four middle sites, and combined two downstream sites.

3. Two groups based on lower vs. higher risk scores for fish populations
4. Five groups based on the location of the sampling sites in the risk regions
5. Eight groups each consisting of one sampling site

The first two categorizations each explained 100% of the variance with their first two canonical discriminant functions, had significant differences between the groups, and demonstrated a high rate of predictability, with the four-group classification correctly identifying 85.7% of the unselected sampling units when using the training set (Figure 6.7). The third classification only explained 56% of the variation and had a low predictive success rate, correctly classifying 76.2% with the training set analysis, only 1.5 times better than random classification. The fourth group explained 100% of the variation, had significant differences between the groups, and demonstrated a high rate of predictability, correctly classifying 71.4% of the unselected cases using the training set, 3.5 times better than random. The fifth set of classifications based on sampling site used four canonical discriminant functions to explain 100% of the variation, but did not show a significant separation for five of the group pairings. This classification also had a low success rate, correctly classifying only 61% of the unselected cases using the training set, although this is still five times better than would be expected from random classification.

The warmwater IBI was run using all fish data identified to species and run separately for each sampling date at each site. The results show most of the sites falling into the fair to good classifications, with the coldwater site (Menges Mill) expectedly falling into the poor classification (Figure 6.8). There were no overall upstream-to-downstream trends apparent and no trends in between-site variance. The uppermost site (Park Road), which showed moderate to extreme differences when compared to the other seven sites in all the other analytical methods, appeared

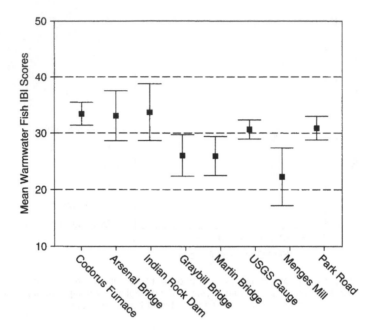

Figure 6.8   Mean scores for warmwater fish IBI ± 2 standard errors.

to be similar to four of the other sites in the IBI. Two of the middle sites (Graybill Bridge and Martin Bridge) did show lower scores than the other sites.

The coldwater IBI (not shown) gave a fair rating to the one true coldwater site (Menges Mill) and rated the other three sites evaluated as poor, indicating they are not true coldwater sites.

Macroinvertebrate Analysis

The dataset for the macroinvertebrate assemblages consisted of 178 categories, most identified to genus with some identified to order or to family. To evaluate trends, we used a subset of 48 macroinvertebrates identified to genus that we found to be significantly nonparametrically correlated with site.

Our PCA on untransformed data identified 12 macroinvertebrate genera (Table 6.3), which allowed us to separate six of the seven sites using the first three components and explained over 58% of the variation. Plotting the first and second components, accounting for over 42% of the variation, showed a separation along two axes. One axis, running from the upper left to lower right, contains all but two of the middle sites (Graybill and Martin Bridge). The second axis, which runs perpendicular to the first, contains Graybill and Martin Bridge (Figure 6.9). The single plots of the factor scores for each site showed no clear upstream-to-downstream trends (not shown).

The clustering analysis had variations in clustering outcome between the different methods used, but similarities appearing across the methods indicated three strong trends, all of which supported the PCA:

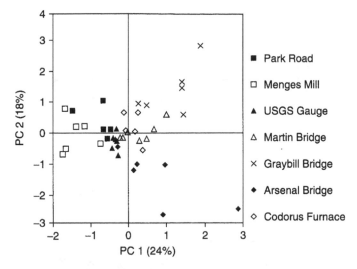

Figure 6.9 Scatter plot of the first and second factor scores from PCA analysis on the macroinvertebrate community data.

1. Two of the inner sites are unique (Graybill and Martin Bridges).
2. The two outermost downstream sites and the three outermost upstream sites grouped together.
3. The two outermost upstream sites showed some separation from the others.

Unlike the fish data, none of the binary clustering methods showed clustering by sites for the macroinvertebrate data.

For discriminant analysis, data for 10 of the 12 macroinvertebrates used in the PCA were transformed using $\log (X + 1)$. For the remaining two genera, one (Diptera *Chironomidae Paratanytarsus*) was left untransformed and another (Trichoptera *Hydroptilidae Hydroptila*) could not be transformed to give a normal distribution and heterogeneity of variance, so was left out of this part of the analysis. The program was asked to classify the sampling units based on seven different groups of categories: the first two based on PCA and clustering results, the third and fourth groups based on clustering results only, the fifth and sixth based on the CCW EcoRA, and one based on sampling site. The classifications were:

1. Two groups, the first containing the two middle sites, Martin and Graybill, the second group consisting of the remainder of the sites
2. Three groups, similar to the first categorization but splitting the second group into one group of the three most upstream sites and another group of the two downstream sites
3. Two groups consisting of one group with the two most upstream sites and another group with the remainder of the sites
4. Three groups consisting of one group with Martin Bridge only, another group with Graybill Bridge only, and a third group of the remainder of the sites
5. Three groups separated by high, medium, or low scores on the CCW EcoRA macroinvertebrate endpoint

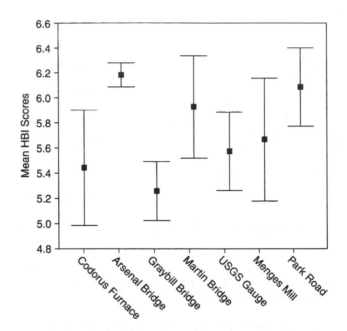

Figure 6.10   Mean scores for macroinvertebrate HBI ± 2 standard errors.

6.   Five groups based on the risk regions
7.   Seven groups each containing one sampling site

The first categorization had significant separation between the groups but had widely differing results between the two methods, with the leave-one-out method having a small eigenvalue of 1.683 and leaving 37% of the variation unexplained, while the training set method had a much larger eigenvalue of 110.738 and left only 1% unexplained. This was a departure from the rest of the analyses, where the differences between methods were small. However, they both successfully classified 88% of the time. The third and fourth sets of categories all had reasonable sized eigenvalues, significant separation of groups, explained most of the variance, and had high rates (>80%) of success in predictive classifications. The fifth categorization was unsuccessful, having very small eigenvalues and leaving 42% of the variance unexplained in the leave-one-out method, and was unable to compute for the training set since no variables qualified for entering or leaving the stepwise method. The sixth and seventh sets of groupings had good success with the leave-one-out method, with high eigenvalues, significant separation of groups, explanation of most of the variation, and correct classification 90% of the time. However, for the training set method, they lost significance between some of the group pairings, with the downstream sites not differing significantly, the downstream and upstream sites not differing significantly, and some of the middle and upstream sites not differing significantly.

The mean HBI scores (Figure 6.10) put all the sites into the polluted category when evaluated using the more stringent criteria (Matthews et al. 1998). When the higher evaluation criteria were used (Snyder et al. 1996), most of the sites were in the middle category of "deviating somewhat from reference value." Three sites (Arsenal Bridge, Martin Bridge, and Park Road) had some scores in the "deviating strongly from reference values" category, although their mean values were in the middle range. The HBI showed no overall upstream-to-downstream trend and no trend in variation.

Combined Analysis

All of the fish and macroinvertebrates that were nonparametrically correlated with site were combined into one dataset consisting of 67 variables. PCA was run on this dataset to determine the best combination of variables that would allow separation of sampling sites with the most variation explained. The resulting PCA consisted of 24 variables, the same 12 fish and 12 macroinvertebrates from the separate analyses. The best PCA result (not shown) explained 49.15% of the variation with the first three components and allowed for clear separation of five of the seven sites (Park Road, Menges Mill, USGS Gauging Station, Martin Bridge, and Graybill Bridge), with the two most downstream sites overlapping (Arsenal Bridge and Codorus Furnace). When the individual factor scores were plotted by site, the first two components (accounting for 37% of the variation) had a clear upstream-to-downstream trend, with the second component showing an additional trend in diminishing variation as you moved upstream.

Clustering was done on the larger combined dataset of 67 variables and on the reduced dataset of 24 variables, with similar results. While there were variations in clustering patterns depending on the method used, there were similarities across methods, indicating strong trends. These trends were:

1. Two middle sites (Graybill and Martin Bridges) are unique.
2. The outer sites (three most upstream and two most downstream) are similar.
3. The two most upstream sites show differing trends.

These clustering results are consistent with the separate fish and macroinvertebrate clustering.

DISCUSSION OF THE CONFIRMATION OF THE RISK ASSESSMENT

We identified four distinct outcomes in the results:

1. We were able to verify the risk hypotheses generated for the biological endpoints, fish and macroinvertebrates, of the CCW EcoRA.
2. We saw reduced variation as an impact in the fish populations.
3. The fish and macroinvertebrate populations showed different responses to the stressors in the system.
4. The biotic indices did not reveal any trends.

The results of the PCA, cluster analysis, and determinants analysis were very similar, enabling us to establish with confidence a set of biological variables that

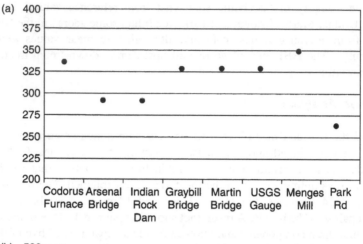

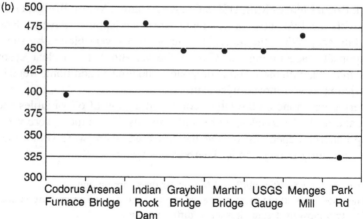

Figure 6.11   Endpoint risk ranks from the CCW EcoRA: (a) ranks for local fish populations, (b) ranks for decline in food availability for aquatic species.

individually characterized the stream sites relative to each other. This individual characterization allowed us to identify trends along the stream and test the risk predictions generated by the CCW EcoRA (Figure 6.11).

The resulting biological variable set (Table 6.3) contained 12 fish with a variety of habitat needs, tolerance levels, and trophic groups (Cooper 1983; Page and Burr 1991). Of importance to resource managers, some of the fish in this group had been previously identified as important to stakeholders. The 12 macroinvertebrates in the variable set were also a diverse group, coming from four orders, with a variety of different habitats and environmental needs and tolerance levels (Thorp and Covich 1991). However, although the group included collector-filterers, collector-gatherers, piercer-herbivores, and scrapers, it omitted predator and shredder functional feeding groups.

Comparing the fish PCA factor scores plotted individually (Figure 6.6) to the EcoRA risk scores for the local fish populations (Figure 6.11a), it is evident that there are similar upstream-to-downstream trends. The CCW EcoRA risk scores showed a trend in increased risk as one moves upstream from Arsenal Bridge to Menges Mill. The PCA plots showed three distinct upstream-to-downstream trends between Arsenal Bridge and Menges Mill. Additionally, the risk scores showed the outside sites were unique from this trend, which is confirmed in the second and third PCA factor scores. There were two additional important trends made apparent in the PCA. First, the reduced within-site variance at the four middle sites indicates that these sites are constrained from normal variation. Second is the inclusion of the site 1 km upstream from the mill effluent outlet in the group of constrained sites, giving evidence for an upstream effect.

The clustering analysis of the fish data confirmed the uniqueness of the outer sites predicted by the EcoRA and the PCA grouping of the four middle sites together.

The discriminant analysis results for the fish demonstrated that the best classification system was a four-group set consisting of the two downstream sites, the four middle sites, the penultimate upstream site, and the most upstream site. This supported the separations seen in the two-dimensional plot of the PCA factor scores and the CCW EcoRA trend. The combining of the two downstream sites, Codorus Furnace and Arsenal Bridge, suggests that a downstream effect might not have been given sufficient weight in the EcoRA. Performing only slightly less successfully was the five-group set based on the location of the sampling sites in the risk regions, providing supportive evidence for the delineation of the risk regions.

Comparing the macroinvertebrate PCA factor scores to the CCW EcoRA risk scores for food for aquatic species (Figure 6.11b), it is evident that none shows a distinct upstream-to-downstream trend; they all share a wave-like pattern. One unexpected finding was a short upstream-to-downstream trend seen in the first PCA factor score between Menges Mill and Graybill. This component had positive loading for all three genera of Coleoptera and both of the genera of Trichoptera with negative loading of four genera of Diptera. This would indicate that along this stretch of the creek the locations upstream of the mill outfall have more of the tolerant Diptera and the locations downstream of the mill outfall have more of the intolerant Coleoptera and Trichoptera.

The clustering analysis of the macroinvertebrate data confirmed the uniqueness of the outer sites as predicted by the CCW EcoRA and the fish analysis, and confirmed the separation of two middle sites (Graybill and Martin) as unique from the rest of the sites.

The discriminant analysis for the macroinvertebrates did not give consistent results for the CCW EcoRA categories, indicating further data are needed to understand what impacts are affecting these groups. One of the uncertainties addressed in the CCW EcoRA was the possibility that the ranks for Regions 1 and 4 might be underestimated, and this may have contributed to the poor match in classification results.

The analysis for the combined dataset of fish and macroinvertebrate was consistent with the separate analyses, but showed no unique trends, indicating that separate analysis of the two populations is sufficient.

The fish and macroinvertebrate IBIs did not give the level of information found in the multivariate analysis. The fish IBI scores provided us with no trends to

evaluate. Using a different IBI for one site (coldwater instead of warmwater) neces-
sitated removal of that site from the overall upstream-to-downstream trend analysis.
Even discounting the coldwater site from the warmwater IBI, there are still no
obvious between-site or within-site trends. The coldwater IBI, with only four metrics,
was likely too general to provide any more detail than confirming that Menges Mill
was a true coldwater site, but the warmwater IBI with 10 metrics should have been
sensitive enough to pick up any trends and reductions in within-site variation (Figure
6.8). The HBI, like the IBIs, did not show any consistent upstream-to-downstream
trends or any within-site trends. The difference between the amount of information
and the detail of information provided by the multivariate analysis as opposed to
the IBIs is significant when viewed from the perspective of a resource manager
needing to make decisions. The multivariate analysis provided specific within-site
population information, within-site level of impact information, and site-to-site trend
information that could be used by the decision maker to change or maintain specific
parts of the system.

A study by Norton et al. (2000) found they were able to separate sites on a high
to low stressor gradient using multivariate statistical analysis (PCA and discriminant
analysis) on datasets constructed of transformed fish and invertebrate IBI metrics.
Our study showed similar abilities to separate sites by using direct analysis of the
fish and macroinvertebrate populations without the need to construct metrics or
transform the data. We argue that these operations, transformation of data and the
construction of metrics, result in the elimination of much of the information that is
contained in the raw data. Our study clearly demonstrates that these operations are
unnecessary. Norton et al. (2000) also found results that suggest the fish and macroinvertebrate communities respond in distinctive and consistent ways to different
types of stress that could be used to build empirical predictive models. Our fish and
macroinvertebrate PCA results are consistent with their findings.

CONCLUSIONS

With this study we were able to establish that the RRM method of EcoRA
produces risk ranking hypotheses that can be tested using multivariate statistical
analysis. Previous studies using the RRM method have demonstrated that it provides
resource managers with risk predictions usable for management decisions (Wiegers
et al. 1998; Obery and Landis, 2002). This study provides evidence of the robustness
of the RRM method of EcoRA and a practical method of testing the risk hypotheses
generated.

We identified areas of uncertainty in the data collection and data analysis phases
of this study that future studies can address, including collection of data in the three
unsampled risk regions to fill in the data gap, standardization of collection methods
for the macroinvertebrates, testing the assumptions made in the data, and testing the
appropriateness of the metrics used in the warmwater and coldwater modified fish
IBIs.

This study also provided evidence of the strong range of detail that can be
provided by the analysis of raw biological data with PCA. There are specific benefits

to be derived from being able to use a few selected fish and macroinvertebrates to characterize a stream. We were able to identify reduced variation as an impact on the fish and we were able to see distinct differences in the responses of the fish and the macroinvertebrates. There are several positive management implications in these abilities. First, we were able to identify several different trends along the creek while avoiding application of labels that imply human value judgments of good or bad, healthy or unhealthy. This allows us to investigate those trends without obvious external biases exerting possible effects on the outcome. Second, unbiased examination of the specific fish and macroinvertebrates allows us to answer directly the needs and wants of the stakeholders that often include a desire for nonnative species or nonhistorical conditions. Third, examination of the trends allows resource managers to target specific areas for change or for maintenance. Fourth, by using a limited number of fish and macroinvertebrates, the analysis supplies resource managers with a monitoring plan with focused endpoints, thus eliminating the collection of unnecessary data, a costly yet common situation.

ACKNOWLEDGMENTS

Thanks to the National Council for Air and Stream Improvements for their financial backing of the project and dedication to the advancement of regional-scale risk assessment. We also thank Leo Bodensteiner and Tim Hall for leading the teams for the collection of the biological datasets used in the confirmatory analysis, and Shawn Boeser, Gene Hoerauf, and Matt Luxon from Western Washington University for their excellent GIS support. Last, we thank the many kind folks in Pennsylvania for sharing historical and current information in the study area.

REFERENCES

Cooper, E.L. 1983. *Fishes of Pennsylvania and the Northeastern United States.* The Pennsylvania State University Press, University Park, PA.

Landis, W.G. and Yu, M-H. 2004. *Introduction to Environmental Toxicology,* 3rd ed., CRC Press, Boca Raton, FL.

Landis, W.G. and Wiegers, J. 1997. Design considerations and a suggested approach for regional and comparative ecological risk assessment, *Hum. Ecol. Risk Assess.,* 3(3), 287–297.

Lyons, J., Wang, L., and Simonson, T.D. 1996. Development and validation of an index of biotic integrity for coldwater streams in Wisconsin, *N. Am. J. Fish. Manage.,* 16, 241–256.

Lyons, J. 1992. Using the Index of Biotic Integrity (IBI) to Measure Environmental Quality in Warmwater Streams of Wisconsin. NC-149:1-51. Technical Report. U.S. Department of Agriculture, Forest Service, St. Paul, MN.

Matthews, R.A., Matthews, G.B., and Landis, W.G. 1998. Application of community level toxicity testing to environmental risk assessment, in *Risk Assessment: Logic and Measurement,* Newman, M.C. and Strojan, C.L., Eds., Ann Arbor Press, Chelsea, MI, 225–253.

Moraes, R., Landis, W.G., and Molander, S. 2002. Regional risk assessment of a Brazilian rain forest reserve, *Hum. Ecol. Risk Assess.,* 8, 1779–1803.

National Council for Air and Stream Improvement (NCASI). 2002. Long-Term Receiving
    Water Study Data Compendium: August 1998 to September 1999. Technical Bulletin
    No. 843, National Council for Air and Stream Improvement, Research Triangle Park,
    NC.
NCASI. 2003. Long-Term Receiving Water Study Data Compendium: September 1999 to
    August 2000. Technical Bulletin No. 856. National Council for Air and Stream
    Improvement, Research Triangle Park, NC.
Norton, S.B., Cormier, S.M., Smith, M., and Jones, R.C. 2000. Can biological assessments
    discriminate among types of stress? A case study from the eastern Corn Belt plains
    ecoregion, *Environ. Toxicol. Chem.*, 19, 1113–1119.
Obery, A.M. and Landis, W.G. 2002. Application of the relative risk model for Codorus Creek
    watershed relative risk assessment with multiple stressors, *Hum. Ecol. Risk Assess.*,
    8, 405–428.
PADEP (Pennsylvania Department of Environmental Protection). 1998. Pennsylvania Code.
    Title 25. Environmental Protection. Chapter 93. Water Quality Standards. Bureau of
    Watershed Conservation. Division of Water Quality Assessment & Standards, Har-
    risburg, PA. Adapted July 18, 1998.
Page, L.M. and Burr, B.M. 1991. *A Field Guide to Freshwater Fishes of North America North
    of Mexico*, Houghton Mifflin, Boston, MA.
Snyder, B.D., Stribling, J.B., and Barbour, M.T. 1996. Codorus Creek Biological Assessment
    in the Vicinity of the P.H. Glatfelter Company, Spring Grove, Pennsylvania. Tetra
    Tech. Inc., Owings Mills, MD.
Sparks, T.H., Scott, W.A., and Clarke, R.T. 1999. Traditional multivariate techniques: potential
    for use in ecotoxicology, *Environ. Toxicol. Chem.*, 18, 128–137.
SRBC (Susquehanna River Basin Commission). 1991. Codorus Creek Priority Water Body
    Survey Report Water Quality Standards Review. Resource Quality Management &
    Protection Division. Publication 134, January.
Thomas, J.T. 2001. An Evaluation of a Relative Risk Model Ecological Risk Assessment in
    Predictive Sustainability Modeling. Master's Thesis. Western Washington University,
    Bellingham, WA.
Thorp, J.H. and Covich, A.P. 1991. *Ecology and Classification of North American Freshwater
    Invertebrates*. Academic Press, New York.
USGS (U.S. Geological Survey). 1999. Occurrence of Organochlorine Compounds in Whole
    Fish Tissue from Streams of the Lower Susquehanna River Basin, Pennsylvania and
    Maryland, 1992. Prepared by Bilger, M.D., Brightbill, R.A., and Campbell, H.L.,
    Lemoyne, PA. Water Resources Investigations. Report 99-4065.
Walker, R., Landis, W.G., and Brown, P. 2001. Developing a regional ecological risk assess-
    ment: a case study of a Tasmanian agricultural catchment, *Hum. Ecol. Risk Assess.*,
    7, 417–439.
Wiegers, J.K., Feder, H.M., Mortensen, L.S., Shaw, D.G., Wilson, V.J., and Landis, W.G.
    1998. A regional multiple-stressor rank-based ecological risk assessment for the fjord
    of Port Valdez, Alaska, *Hum. Ecol. Risk Assess.*, 4, 1125–1173.

CHAPTER 7

Codorus Creek: Use of the Relative Risk Model Ecological Risk Assessment as a Predictive Model for Decision Making

Jill F. Thomas

CONTENTS

1-56670-655-6/04/$0.00+$1.50
© 2004 by CRC Press LLC

INTRODUCTION

Sustainability of resources requires decision-making tools that predict how different management options will impact multiple aspects of an ecosystem, allowing for decisions that optimize results while minimizing risk. The conceptual model developed as part of the relative risk model (RRM) of ecological risk assessment (EcoRA) can be used as such a predictive tool. The conceptual model developed for the Codorus Creek Watershed EcoRA was tested for predictive modeling. The process involved five steps: defining decision options, determining impacted sources of stress and receptors, calculating the change in rank for each impacted source and receptor, calculating the change in endpoint risk scores, and analyzing the predicted change in risk patterns. Decision options that were tested included positive actions, negative actions, or no action. For each tested decision option the conceptual model provided a prediction of probable changed pattern of risk for each endpoint across the projected impacted risk regions. The results indicated most options resulted in subtle changes in the watershed; there was no one overall decision that would reduce risk for the entire watershed. These results demonstrate that the RRM conceptual model is easily used as a decision-making tool, providing clear usable information.

DECISION MAKING, SUSTAINABILITY, ADAPTIVE MANAGEMENT

Resource managers have a number of needs left unmet by the current options of decision-making tools. Management of resources necessarily involves making assumptions about both future conditions of the ecosystem and about the expected impact of management actions. Few techniques, if any, deal with the predictive nature of management. Additionally, management goals often include statements about recovery or sustainability, two terms with lacking widely accepted operationally defined meanings.

One of the most widely accepted methods currently in use for managing resources is adaptive management. The drawbacks of this method are the requirement for large amounts of data and an extensive timeline to test the initial outcomes before feedback enters the decision-making process. Few attempts at this method have met both these requirements, and so it remains widely used but substantially unproven (Lee 1999). Additionally, adaptive management does not have any mechanism for *a priori* comparison of alternative decision options, resulting in a trial-and-error method of management.

The use of the term *recovery* as a management goal needs to be replaced with a concept that can be operationally defined. Recovery to a preexisting condition prior to European settler impact is an unreasonable goal and should be ruled out. Recovery as defined by a reference site is also unreasonable given that no two sites have identical histories and so will always have inherent differences. Recovery as defined by preset static points for indicator species ignores the dynamic nature of ecosystems. A preferable goal for management would be directionality of movement of measurement endpoints within defined limits set by stakeholders as proposed by Landis and McLaughlin (2000a). This is achievable, easily defined, targeted to management and stakeholder goals, and requires relatively few measurements over time for verification.

EcoRAs are in a unique position to assess potential changes in the ecosystem. The risk characterization step in the USEPA Guidelines (1998) addresses the future state of a site using the concept of recovery and defining it as reversible vs. irreversible changes to structural or functional components in the ecosystem. However, a drawback to the USEPA format is that it does not incorporate a standard methodology for evaluating future trends in risk. A recent issue of *Human and Ecological Risk Assessment* (April 2001) featured a Debate and Commentary section on regional-scale ecological assessment of cumulative risks in which several of the commentators remarked on the need for risk assessments to include predictive decision-making aspects (Moore 2001; Gentile and Harwell 2001). Moore (2001) argued for an EcoRA method that includes societal buy-in, broad inclusion of stressors and options, use of modeling, and an adaptive management style of action, observation, and revision. Gentile and Harwell (2001) evaluated a number of different risk assessment methods and argued in favor of developing a common approach based on the USEPA Guidelines (1998) that would include multiple ecological components, interaction between components, societal goals, and the ability to predict future risks.

The RRM method of risk assessment (Landis and Wiegers 1997) provides societal buy-in through stakeholder-derived endpoints, maps multiple stressors and receptors in their geographic context, and creates a conceptual model of interactions, all aspects highlighted as necessary by both Moore (2001) and Gentile and Harwell (2001). The conceptual model generated by the RRM EcoRA has built-in flexibility that allows for easy addition, deletion, or modification of components (stressors, receptors, or endpoints) or structural features (pathways). This flexibility of the model allows it to remain current as changes occur in the ecosystem (in response to decision actions or natural impacts) or as additional data are obtained. This attribute fulfills the need for the adaptive management type of response for evaluating outcomes of decisions that Moore cited as important. An additional necessary criterion not discussed by the commentators is testability of the model. An EcoRA that provides testable hypotheses allows for confirmation of the risk predictions. The RRM format allows for the creation of hypotheses in the form of patterns of predicted risk. These patterns can easily be tested to confirm the risk assessment. A confirmed EcoRA provides us with the ecological position (Landis and McLaughlin 2000b) of each endpoint in relation to the stakeholder set limits. This then gives the resource manager the ability to identify those elements that are outside or close to the limits and therefore are more likely to need management action. Additionally, a risk assessment that has been confirmed in a previous step will carry more weight in the decision-making step.

RELATIVE RISK MODEL EcoRA

The RRM EcoRA has been used in multiple settings (Wiegers et al. 1998; Luxon 2000; Walker et al. 2001; Obery and Landis 2002) and been proven adaptable to meet the needs of each situation. Three components of the RRM method make it uniquely suited to predictability modeling. First is its ability to evaluate multiple chemical and physical stressors and multiple endpoints. This allows the RRM to be applicable in the widest possible number of areas. Second is the generation of testable hypotheses in the form of relative magnitude and absolute patterns of risks, allowing for confirmation of the model. The ability to test and verify the RRM EcoRA provides users with confidence in the results and credibility when making recommendations based on the patterns of risk. Often, management decisions are made from choices generated by conflicting stakeholder values, and the decisions may be controversial and may involve public scrutiny. In these cases, the use of a confirmed model with probable predictions would give the decision makers a tool to test and explain their decisions. An additional advantageous outcome of the hypothesis testing is the generation of specific endpoints that can be used in a monitoring program to evaluate future changes. The third component of the RRM EcoRA is its ease of use and clarity of output as a decision-making tool. This provides resource managers with clear information on the areas and endpoints that are the most probable to be moved in the desired direction, allowing them to focus their management efforts.

The RRM EcoRA developed for the Codorus Creek Watershed in south central Pennsylvania, covered in Chapter 6, was selected for use as a predictive model for two reasons. First, Codorus Creek is a heavily impacted waterway with multiple stressors in a watershed with a long history of monitoring. Second, the RRM EcoRA risk predictions were confirmed for two biological endpoints, fish and macroinvertebrates (Thomas 2001).

The Codorus Creek EcoRA was derived entirely from existing data collected by local, state, and federal agencies and organizations. A conceptual model was created using geographic information systems (GIS) to break down the watershed into smaller risk regions and categorize stressors, habitats, and complete pathways. Stressors were identified as landuse, soil erosion, surface runoff, streambank development, illegal waste disposal, wastewater discharges, altered flow rates, and altered channel structure. Habitats of interest were identified as macroinvertebrate, fish, riparian, and urban park. Effects were assessed by assigning high, medium, and low ranks to the stressors and habitats. Risk characterization ranked complete exposure pathways established in the conceptual model, and relative ranks were summed for all the sources in each risk region, providing a relative risk per endpoint and an overall relative risk per risk region. Verification of the risk scores for the fish population and macroinvertebrate population endpoints was achieved by applying three multi-variate statistical methods (principal components analysis, hierarchical clustering, and discriminant analysis) to an independently collected fish and macroinvertebrate assemblage dataset and comparing the resulting patterns to the patterns of risk generated by the EcoRA. The patterns bore strong resemblance in both upstream-to-downstream gradients and in outliers to the predicted risk assessment (Spromberg et al. 1998).

PURPOSE OF STUDY AND SUMMARY OF RESULTS

The purpose of this study was to test the Codorus Creek RRM EcoRA as a predictive model for decision making. In the process of carrying out this study we addressed three main questions:

1. Does the model provide clear indications of direction of movement of risk for specific endpoints?
2. What are the uncertainties in the method and what measures can be taken to reduce those uncertainties?
3. Based on the predictions of the EcoRA does the conceptual model need refinement?

We tested the predictive methodology by evaluating different management decision options and comparing their probable impacts on the watershed. Our results demonstrate that the RRM is easily used as a decision-making tool, providing clear, usable information. Resource managers can use this information to select the decision option that will give the highest probability of the desired outcome while minimizing the risks and the costs. We identified a number of areas of uncertainty and how those uncertainties can be reduced. Finally, we identified areas of the conceptual model that need to be refined in order to more accurately predict changes in the watershed.

METHODS

Codorus Creek RRM EcoRA Methods

The Codorus Creek EcoRA is detailed in Chapter 6. The risk hypotheses generated by the landscape pattern of risk for two biological endpoints, fish and macroinvertebrates, were tested using multivariate statistics on an independent dataset, with substantial pattern similarities found in the existing aquatic communities to the predicted endpoint risk patterns, thereby confirming the Codorus Creek RRM EcoRA (USEPA 1998). With the current endpoint positions confirmed, future directionality and magnitude of movement can be assessed by manipulation of the source and habitat ranks based on projected changes resulting from actions taken in the watershed and comparing the result to the original endpoint risk scores.

RRM EcoRA Predictive Model Method

There are five steps involved in the use of the RRM EcoRA as a predictive model:

1. Clearly defining the decision option to be assessed
2. Determining all sources and habitats that would be impacted by this decision option
3. Calculating the change in rank for each impacted source and receptor, using the original risk assessment criteria
4. Calculating the change in endpoint risk scores
5. Analyzing the predicted change in risk patterns

The first step, clearly defining the decision option, requires a detailed description of what the decision option entails, a complete list of all the assumptions that are being made with this option, consideration of the spatial scale of the potential changes due to the option, and an explanation of how effects of the option are to be input into the conceptual site model. The goal is to make the process as clear as possible in order to reduce any miscommunication on what is meant by the decision option and to make clear all the related uncertainties involved in the predictive process.

The second step, determining all impacted sources and habitats, needs to consider the ramifications of the decision option on the sources, such as changes in landuse or stream characteristics, and on the receptor habitats, such as changes in area or quality of habitat.

The third step in the process, calculating the change in rank for sources and receptors, applies the changes determined in the second step to the original calculations of rank for each of the affected features. In order to determine directionality of movement in the final step, it is critical to use the original EcoRA criteria for determining rank in this step.

The fourth step, calculating the change in endpoint risk scores, enters the new ranks in the conceptual site model spreadsheet to recalculate the endpoint risk scores. The final step plots the percent change in endpoint scores to evaluate the probable movement and direction of the endpoint in response to the decision option. The outcomes for several alternate decision options can be plotted together to evaluate which option results in the best outcome.

Tested Decision Options

For this study six potential decision options were selected. These choices were made based on discussions with stakeholders, the results of the Codorus Creek RRM EcoRA, and our desire to evaluate the outcome to extreme measures.

1. Increase the forested area in the riparian corridor
    The intent for this option is to decrease the risk in the three risk regions most heavily impacted by industrial and urban landuse. The method by which this is to be done is by improving the quality of the land in the riparian corridor; as such, the riparian corridor is not expanded in this option, but is improved by converting 10% of the nonforested riparian land into forested land. Reforesting of riparian habitats is a popular and visible stream rehabilitation method; this option will test how this activity is predicted to alter the risk patterns for these regions.

2. No action, resulting in a 10% increase in urban landuse
    The purpose of this action was to assess the movement for the endpoints to expanding urbanization. Several of the risk regions that have excellent riparian corridors have development taking place close to Codorus Creek. This option will test what would happen to the endpoint risk patterns if the development in the watershed were unchecked. The expansion was set at a 10% increase in urban landuse to reflect the predicted 25-year population growth for this area; additionally, the assumption was made that the increased urban area would come out of land currently in agricultural usage. As region 2 is entirely urban, this region was not included in this option.

3. A 50% reduction in effluent constituents from a major industrial source
   Two concerns voiced by the stakeholders about the water in Codorus Creek were objections to the color and odor they thought came from effluent emanating from a pulp and paper mill located in risk region 3. In response to this, the mill is currently undergoing modifications to reduce the effluent color and odor constituents by 50%. This option will test the potential response in risk patterns to this change. The assumption was made that reducing the color and odor constituents by 50% would also reduce other relevant chemical constituents by 50%.

4. Diverting 50% of storm runoff into treatment facilities
   The segment of Codorus Creek that runs through the City of York is heavily channelized with many drains opening into the channel. Some of these are for untreated stormwater runoff. This option tests the risk pattern response to reduction of untreated stormwater runoff by 50%. The assumption was made that all risk regions could divert stormwater runoff to treatment facilities.

5. Elimination of a major industrial source
   One of the most extreme measures that could be taken would be to remove the pulp and paper mill in region 3. This option tested how this drastic alteration would change the risk patterns in the watershed.

6. Reduction of agricultural impact
   The Codorus Creek RRM EcoRA determined that the largest contributor to risk was agricultural landuse. This option tested how a 50% reduction in agricultural stressors would change the risk to the affected areas. The reduced stressor load could be from either improved agricultural methods (e.g., use of best management practices such as alternatives to chemical pesticides and containment of wastes) or from reduced agricultural landuse. However, for the model, reduced agricultural landuse was used to simulate either reduction method. Landuse areas were not changed in response for any other landuse component.

Uncertainties

Uncertainties in the risk predictions for the original EcoRA are covered in Chapter 6. In the information used for the predictive modeling, the uncertainties stemming from lack of knowledge may have resulted in errors. Included in this are:

- The Codorus Creek predictive model does not have a temporal component; therefore, the predictions for all the endpoints are uncertain due to lack of knowledge about the trajectory of their current movement. If the true trajectory is opposite to the predicted movement, the final outcome could be a counterbalancing of the two directions resulting in a slowing down or halt of movement in that opposite direction, giving the impression of an error in the prediction.
- The predictions do not account for patch dynamics, so "action at a distance" (Spromberg et al. 1998; McLaughlin and Landis 2000) may occur, resulting in a buffering effect or an exaggeration on the predicted movement, giving the appearance of a slowing down or nonmovement of the endpoint (Landis and McLaughlin 2000a; 2000b).
- The predictions for the unverified endpoints can only give direction and magnitude, but because their ecological position has not been determined, their relative risk of moving out of compliance is unknown.

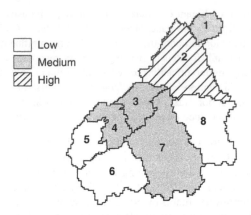

Figure 7.1 Original risk assessment landscape risk pattern.

- The relative stochastic and deterministic components of the Codorus Creek eco-system are unknown; therefore, there is a level of uncertainty on how the com-plexities of the ecological system may result in different outcomes than predicted.
- Magnitude of change is assumed to be equal between the different risk regions. This may not be true, resulting in different patterns of risk change than predicted.

RESULTS

Four of the six options resulted in changes in risk scores from the original risk assessment: options 1, 2, 4, and 6. Two of these (options 1 and 2) had sufficiently large changes to alter the overall landscape risk pattern from the original pattern (Figure 7.1). The results for all the options are described below.

Predictions of Risk Trend Changes for Option 1 (Increase in Riparian Forestation)

This option improves the riparian habitat, resulting in an increased risk rank for one habitat (riparian) with an offsetting decrease in one stressor (streambank devel-opment). These alterations resulted in changed risk scores for regions 2, 3, and 4 (Table 7.1). Risk region 2, the most downstream and most urban of these three risk regions, showed decreasing risk scores for all the endpoints except the fish popula-tion, which increased slightly. Risk regions 3 and 4 had increases in the same three endpoints (water quality, fish population, and stormwater control) with no changes in the remaining endpoints. The most severe impact to an endpoint appears to be predicted for stormwater control for regions 3 and 4, with predicted risks increasing 50% or more. However, this increase is driven by the extremely low risk scores in the original EcoRA, resulting in a small change appearing to be very significant. The increasing risk scores for water quality, fish population, and stormwater control in the two upstream regions (3 and 4) are driven by their low habitat scores in the original risk assessment. Increased risk in this case can be interpreted as improved

Table 7.1 Summary of Percent Changes in Total Risk Scores for Option #1 (increasing forested area in riparian corridor) by Risk Region and Endpoint

Risk Region	Water Quality	Water Supply	Fish Population	Macroinvertebrate Population	Recreational Use	Stormwater Treatment
2	−2.3	−10.0	4.1	−10.0	−15.8	−4.5
3	4.8		7.3			50.0
4	4.0		5.7			55.6

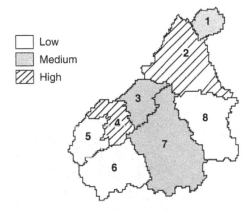

Low

Medium

High

Figure 7.2   Risk pattern for option #1 (increase riparian forestation) showing region 4 with high risk.

habitat. The decreasing risk scores for the mostly urban region 2 are driven by the stressor (streambank development) decreasing more than the offsetting increase in riparian habitat rank.

The total risk score changes resulted in a change in the landscape risk pattern, with region 4 increasing from medium to high risk (Figure 7.2).

Predictions of Risk Trend Changes for Option 2 (10% Increase in Urban Area)

This decision option resulted in impact to six of the nine stressors (agricultural landuse, urban landuse, industrial landuse, soil erosion, streambank development, and runoff), although soil erosion changes were so small they did not change the ranking for this source in any of the risk regions. Almost all of the changes in rank for the stressors were increases; most increasing by 2, but both runoff and urban landuse increased by 4. The one stressor that decreased in rank was streambank development in region 4; the borderline high ranking it had in the original EcoRA drove this drop. Four of the five habitats were impacted (macroinvertebrate, cold-water fish, warmwater fish, and riparian habitat), but only the last two had changes significantly large enough to change their ranking, decreasing by two in regions 1, 4, 6, 7, and 8.

Table 7.2   Summary of Percent Changes in Total Risk Scores for Option #2 (10% increase in urban landuse) by Risk Region and Endpoint

Risk Region	Water Quality	Water Supply	Fish Population	Macroinvertebrate Population	Recreational Use	Stormwater Treatment
1	2.7	9.1		9.1	7.1	−16.7
3	22.6	21.4	23.2	21.4	21.4	33.3
4	2.4	2.6	−3.5	2.6	−3.8	
5	13.9	14.3	13.6	14.3	14.3	10.0
6	23.7	33.3	19.7	33.3	28.6	−10.7
7	7.5	15.4	3.7	15.4	15.4	−43.8
8	29.6	40.0	24.1	40.0	35.7	5.9

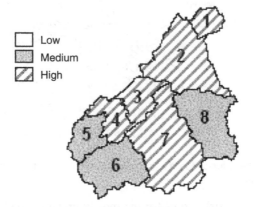

Figure 7.3   Risk pattern for option #2 (10% increase in urban landuse).

These alterations resulted in a change in risk scores for all the targeted risk regions (Table 7.2); region 2 was excluded from changes as it is already predominantly urban landuse. The predicted changes were primarily increases, driven by the overwhelming increases in risk to the stressors. Three endpoints, water quality, water supply, and macroinvertebrates, had increased risk in all seven impacted regions. Two other endpoints, fish population and recreational use, had increased risk predicted in six of the seven regions, both showing slight decreases in risk for region 4, which was driven by the streambank development decrease in rank. Stormwater control showed mixed results, with decreases in predicted risk for regions 1, 6, and 7 and increases in risk for regions 3, 5, and 8. The decreasing risk results were driven primarily by the decrease in riparian habitat in risk regions that originally had high-quality habitat regions, and the increasing risk resulted in increased rank for urban landuse and runoff with no offsetting increases in habitat ranks.

The reductions in risk for the seven impacted regions resulted in an increase in overall risk category for all affected regions (Figure 7.3). Regions 1, 3, 4, and 7 all increased from medium risk to high risk. Regions 5, 6, and 8 increased from low to medium risk.

Table 7.3   Summary of Percent Changes in Total Risk Scores for Option #4
           (50% Increase in runoff treatment) by Risk Region and Endpoint

Risk Region	Water Quality	Water Supply	Fish Population	Macroinvertebrate Population	Recreational Use	Stormwater Treatment
3	−6	−7	−6	−7	−7	
4	−7	−8	−7	−8	−8	
7	−7	−8	−6	−8	−8	

Predictions of Risk Trend Changes for Option 3 (50% Reduction in Effluent Constituents)

The analysis of this potential action resulted is no predicted risk change for any of the endpoints in any of the risk regions at or downstream from the pulp and paper mill. The outcome for this option was driven by only one stressor calculation being changed, wastewater discharges. This stressor was calculated by averaging a rank for amount per day of wastewater discharges with a rank for the number of analytes in the wastewater that exceeded a hazard quotient (HQ) of 1. Removing additional chemical constituents from the wastewater would not reduce the volume of wastewater, but would reduce the number of analytes with HQs greater than 1. Even with the HQ rank reduced by half, the averaged rank remained unchanged for risk region 3 and the two regions downstream. The possible reason for this apparent insensitivity of the model to the reduced HQs is discussed later.

Predictions of Risk Trends for Option 4 (50% Increase in Runoff Treatment)

The risk changes for this option were driven by having no impacted habitat and only one impacted stressor (runoff) with reduced risk in regions 3, 4, and 7. Therefore, all of the endpoints showed either decreased or no change in risk. One endpoint, stormwater treatment, surprisingly showed no change in risk. The other five endpoints all showed small (6 to 8%) reductions in risk for regions 3, 4, and 7 (Table 7.3). These small reductions in risk did not change the overall categories of risk for the three regions in question, resulting in no change in the landscape risk pattern.

Predictions of Risk Trend Changes for Option 5 (Elimination of Mill)

The analysis of this potential action resulted in no predicted risk change for any of the endpoints in any of the risk regions at or downstream from the pulp and paper mill. Only two of the nine stressor calculations were impacted (wastewater discharges and altered stream flow rates), and only two habitat calculations were affected (coldwater fish and warmwater fish). This surprising outcome of unchanged risk patterns was driven by the small changes in the input numbers used to calculate the ranks, resulting in no changes in risk ranks for any of the impacted stressors or habitats.

Table 7.4 Summary of Percent Changes in Total Risk Scores for Option #6
(50% reduction in agricultural stressors) by Risk Region and Endpoint

Risk Region	Water Quality	Water Supply	Fish Population	Macroinvertebrate Population	Recreational Use	Stormwater Treatment
1	−10	−9	−11	−9	−10	−14
7	−8	−8	−9	−8	−8	−12

Predictions of Risk Trend Changes for Option 6 (50% Reduction in Agricultural Stressors)

This decision option reduces the predicted risk by 8 to 14% for all of the endpoints in two risk regions, 1 and 7 (Table 7.4). This reduction is driven by these two regions having the highest percentage of agricultural landuse, 73 and 80%, respectively. No other stressor calculations were impacted, and no habitat calculations were impacted. The reductions in risk did not change the overall categories of risk for regions 1 and 7, resulting in no change in the overall landscape risk pattern.

Sensitivity Testing

To test the sensitivity of these calculations, decision option 6 was evaluated for ranges of outcomes depending on the input. For risk region 1, which had the risk lowered from high (6) to medium (4) by the 50% reduction in agricultural impact, the same result was achieved for agricultural reductions ranging from 10 to 65%. The extreme low end was due to the location of the original landuse percentage near the breakpoint for the high and medium ranks, so a small drop in landuse would result in a reduction in rank. At the other end of the scale, risk regions 3, 4, and 6 were all just below the breakpoint between high and medium and would require much larger reductions in landuse to result in a change in rank; slightly greater than 60% reduction was needed to produce a reduced risk ranking from medium (4) to low (2) for region 6. This indicates the location of the stressor or habitat near to a ranking breakpoint makes it more sensitive to small changes. This effect was also seen in the ranking drop for streambank development in option 2 and the lack of response to reduced HQs in option 3.

DISCUSSION

In the first option the proposed addition of forested riparian habitat demonstrated the increased risk that occurs in an ecosystem when new habitat is generated. This type of result illustrates the need for careful planning when the restoration or creation of new or additional habitat is proposed to ensure that it is suitably shielded from stressors that occur in that risk region. Without proper planning, habitat can put receptors at increased risk of exposure and ultimately work against increasing population size by acting as a sink for any sensitive receptors that rely on this habitat. The decreasing risk scores for the mostly urban region 2 indicate that if resources are limited, then increasing riparian area in region 2 would provide the most benefit with the least amount of risk.

The no-action option resulting in increased urbanization gave the most dramatic increase in risk to the watershed, indicating that zoning and landuse have the biggest potential for impacting the area. Even the few endpoints that had decreasing risk were due to loss of quality habitat. This type of information could indicate to stakeholders that their most productive course to protect the watershed would be to direct their focus on the decisions being made in zoning and landuse offices.

The third and fifth options, each involving reducing or removing the pulp and paper mill impacts, produced the most surprising results, with no change in predicted risk. This should not be taken as proof of negligible impact on the watershed by the mill, but may possibly indicate the need for refinements in the conceptual model. For example, the third option of a 50% reduction in effluent constituents appears to be driven by missing components in the ranking calculations for the sources of stress and for habitats. First, in the sources, is the lack of a component in the wastewater effluent that includes the color and odor of the effluent. Stakeholders had told our group in meetings that their concern over the color and odor of the creek reduced recreational use downstream from the mill. Therefore, the conceptual site model should reflect decreased risk numbers for the recreational use endpoint with a reduction in mill effluent constituents, but this did not occur. The aesthetic quality, odor and color, of the effluent could be measured and included in the wastewater rank calculations. This refinement was tested by changing the wastewater rank calculation to an average of three components, adding an aesthetic color and odor rank to the currently used volume rank and analytes rank. However, even with an original assumed rank of 6 for the aesthetic component being reduced to 4, the risk remained unchanged, indicating these components may need to be treated separately in order to give them additional weight in the final ranking. Second, in the habitats, there is the lack of a component of effluent constituents in the habitat ranking calculations. Each potentially impacted habitat could include this information; the habitat suitability index (HSI) for the fish habitats could be expanded to include chemical constituents; the macroinvertebrate habitat knowledge could be increased to allow for specific calculations of rank; and the urban park habitat rank could be changed from a simple area calculation to an average of area and an aesthetic color and odor component. These changes would produce a more accurate prediction of impact.

The removal of the mill could be expected to result in a number of dramatic changes to the watershed, and this should be reflected in the predicted risk patterns. The lack of change indicates missing components or missing pathways. The aesthetic and constituent components discussed above would result in some level of change. Another possible missing component is landuse change; although it is unlikely that the industrial site would revert to wooded riparian habitat, it might not remain industrial. Fish and macroinvertebrate habitat changes from the removal of wastewater from the stream would be expected, yet this did not appear in the calculations. Increased knowledge of fish and macroinvertebrate habitat would allow for refinement of these ranks.

The fourth option of diverting storm runoff into treatment facilities also produced a surprising result with the one endpoint that would be expected to change (stormwater control) showing no predicted change in risk. Analysis of the conceptual site model showed no link between runoff and stormwater control. Addition of this expected complete pathway would result in more realistic results.

One of the assumptions discussed above in the RRM uncertainties is the assumption that an increased quantity of a stressor results in increased risk. The final option, reducing the impact from agriculture, demonstrated this effect on the outcome of the prediction process. Additionally, this option demonstrated the importance of the ecological position of the stressor relative to the natural breaks used to divide the quantities into high, medium, and low categories, which strongly affected the outcome of the predictive process.

CONCLUSIONS

These results support the ability to use the RRM EcoRA as a predictive decision-making tool. The conceptual model provided transparent, easily understood movement of the endpoints in response to the decision option parameters. The identified uncertainties associated with using this model, which include the uncertainties from the original risk assessment, the confirmation process that determines ecological position of the endpoints, the uncertainties from lack of knowledge, and the uncertainties due to stochastic processes, provide a margin of error in use of the predictions. Additionally, the process provided a feedback loop for refinement to the original conceptual model, which allows for more accurate and detailed responses to be predicted.

The results demonstrate that there is no one management action that will reduce risk for all the endpoints. There are a number of reasons for this. First, the endpoints are derived from stakeholder values and so may have conflicting needs not only between endpoints, but also within endpoints. An example of a conflict within an endpoint is the Codorus Creek fish population endpoint — the stakeholders value both warmwater and coldwater fish. In order to manage for both these types of fish, the resource manager must maintain the islands of cold and warm water that are the result of the hypolimnetic coldwater input from an upstream dam for the coldwater habitat and the warmwater effluent discharge from industry for the warmwater habitat. This results in conflicted outcomes where an improved condition for one endpoint may result in a decline for another. Second, the commonly held concept that improving habitat reduces risk is not valid. Improved or increased habitat will, in fact, be at more risk from stressors, both because new areas are now being exposed to stressors and because opening up new habitat areas may potentially act as islands, removing needed numbers of organisms from the main populations, resulting in an overall decline in population. As was shown in the results of this study, this increased risk in habitat ranking can offset decreased risk from stressors, resulting in changes that are slight and multidirectional along the watershed. Third, this model includes the cascading effect of a change that impacts multiple source–habitat pathways in a region-specific manner, resulting in spatial variations of impact. Fourth, by including this spatial component in the watershed analysis, management actions can be evaluated specifically to the areas they will impact, and any assumption of blanket effect is eliminated.

The current trend in resource management is the use of the single indicator as a means of evaluating the state of the ecosystem and regulating its future. One important conclusion that can be drawn from the subtle variations in responses that resulted from these scenarios is that no one indicator can provide a complete picture of the current state of a system. A single endpoint indicator reflects a narrow picture of the state of the ecosystem based on the values of the stakeholders who generated that endpoint. Additionally, no single measurement point, whether it is an organism or a total maximum daily load (TMDL), can effectively predict or regulate the future state of the system, since it is a projection from a single point, which greatly decreases the probability of accuracy when applied to the larger scale of the watershed with multiple stressors and receptors. The results from this study support the concept that due to the complex nature of ecosystems, a more realistic evaluation must include multiple endpoints from a representative stakeholder group and that these endpoints must be evaluated using watershed-wide information at as many ecological and spatial levels as possible. This model and these results concur with and meet the criteria discussed by Moore (2001) and Gentile and Harwell (2001).

ACKNOWLEDGMENTS

I would like to thank the National Council for Air and Stream Improvement for their support of this work.

REFERENCES

Gentile, J.H. and Harwell, M.A. 2001. Strategies for assessing cumulative ecological risks, *Hum. Ecol. Risk Assess.*, 7, 239–246.

Landis, W.G. and McLaughlin, J.F. 2000a. If not recovery, then what?, in *Environmental Toxicology and Risk Assessment: Science, Policy and Standardization – Implications for Environmental Decisions*, Tenth Volume, ASTM STP 1403, Greenburg, R.N. et al., Eds., American Society for Testing and Materials, West Conshohocken, PA.

Landis, W.G. and McLaughlin, J.F. 2000b. Design criteria and derivation of indicators for ecological position, direction, and risk, *Environ. Toxicol. Chem.*. 19, 1059–1065.

Landis, W.G. and Wiegers, J.K. 1997. Design considerations and a suggested approach for regional and comparative ecological risk assessment, *Hum. Ecol. Risk Assess.*, 3, 287–297.

Lee, K.N. 1999. Appraising adaptive management, *Conserv. Ecol.*, 2, 3. http://www.con-secol.org/vol3/iss2/art3/html.

Luxon, M. 2000. Application of the Relative Risk Model for Regional Ecological Risk Assessment to the Upper Willamette River and Lower McKenzie River, Oregon. Master's Thesis, Western Washington University, Bellingham.

McLaughlin, J.F. and Landis, W.G. 2000. Effects of environmental contaminants in spatially structured environments, in *Environmental Contaminants in Terrestrial Vertebrates: Effects on Population, Communities, and Ecosystems*, Albers, P.H. et al., Eds., Society of Environmental Toxicology and Chemistry, Pensacola, FL.

Moore, D.R.J. 2001. The *Anna Karenina* principle applied to ecological risk assessments of
    multiple stressors, *Hum. Ecol. Risk Assess.*, 7, 231–237.
Obery, A.M. and Landis, W.G. 2002. Application of the relative risk model for Codorus Creek
    watershed relative risk assessment with multiple stressors, *Hum. Ecol. Risk Assess.*,
    8(2), 405–428.
Spromberg, J.A., John, B.M., and Landis, W.G. 1998. Metapopulation dynamics: indirect
    effects and multiple distinct outcomes in ecological risk assessment, *Environ. Toxicol.
    Chem.*, 17, 1640–1649.
Thomas, J.F. 2001. Confirmation of a Relative Risk Model Ecological Risk Assessment of
    Multiple Stressors Using Multivariate Statistics, Master's Thesis, Western Washington
    University, Bellingham.
USEPA (U.S. Environmental Protection Agency). 1998. Guidelines for Ecological Risk
    Assessment. EPA/630/R-95/002F. U.S. Environmental Protection Agency, Washing-
    ton, D.C.
Walker, R., Landis, W., and Brown, P. 2001. Developing a regional ecological risk assessment:
    a case study of a Tasmanian agricultural catchment, *Hum. Ecol. Risk Assess.*, 7,
    417–439.
Wiegers, J.K., Feder, H.M., Mortensen, L.S. et al. 1998. A regional multiple-stressor rank-
    based ecological risk assessment for the fjord of Port Valdez, Alaska, *Hum. Ecol.
    Risk Assess.*, 4, 1125–1173.

CHAPTER 8

Developing a Regional Ecological Risk Assessment: A Case Study of a Tasmanian Agricultural Catchment*

Rachel Walker, Wayne G. Landis, and Philip Brown

CONTENTS

* Previously published in the *Journal of Human and Ecological Risk Assessment,* 7(2), 2001. Reprinted with permission.

INTRODUCTION

A regional ecological risk assessment was conducted for the Mountain River catchment in Tasmania, Australia. The relative risk model (RRM) was used in conjunction with geographic information systems (GIS) interpretations. Stakeholder values were used to develop assessment endpoints, and regional stressors and habitats were identified. The risk hypotheses expressed in the conceptual model were that agriculture and land clearing for rural residential development are producing multiple stressors that have potential for contamination of local water bodies, eutrophication, changes in hydrology, reduction in the habitat of native flora and fauna, reductions in populations of beneficial insects in agricultural production systems, increased weed competition in pastures, and loss of aesthetic value in residential areas. In the risk analysis the catchment was divided into risk regions based on topography and landuse. Stressors were ranked on likelihood of occurrence, while habitats were ranked on percentage of land area. Risk characterization showed risks to the maintenance of productive primary industries were highest across all risk regions, followed by maintenance of a good residential environment and maintenance of fish populations. Sensitivity analysis was conducted to show the variability in risk outcomes stemming from uncertainty about stressors and habitats. Outcomes from this assessment provide a basis for planning regional environmental monitoring programs.

THE RELATIVE RISK MODEL

Regional ecological risk assessment is concerned with describing and estimating risks to environmental resources at the regional scale or risks resulting from regional-scale pollution and physical disturbance (Hunsaker et al. 1989). Within any catchment region there are various stressors impinging on the quality of the environment. Without a framework it is difficult to objectively assess the risks associated with multiple stressors. The RRM as developed by Landis and Wiegers (1997) is a framework for ranking and comparing the risks associated with multiple stressors. It is a useful tool for describing and comparing risks to valued resources within a catchment.

The RRM was developed for a regional risk assessment for the fjord of Port Valdez, Alaska (Wiegers et al. 1998). This chapter reports on the application of the RRM in a more localized region in southern Tasmania, Australia. The aim of this work was to use the RRM as a tool to put catchment issues in context and highlight issues that needed to be further addressed.

RRM methodology essentially mirrors the traditional three-phase risk assessment approach: problem formulation, analysis, and risk characterization, but requires a modification of the traditional approach. Expanding an assessment to cover a region requires consideration of larger scale, regional components: sources that release stressors, habitats where the receptors live, and impacts to the assessment endpoints.

In the problem formulation phase of the relative risk assessment, the scope of the assessment is defined; at this stage the values of regional stakeholders are influential in determining assessment endpoints. Generic goals for regional risk assessment include explanation of observed regional effects, evaluation of an action with regional implications, and evaluation of the state of a region (Suter 1990). Regional stressors and habitats are identified in the problem formulation phase.

In the risk analysis phase the stressors and habitats are ranked based on their likelihood of occurrence within the risk region. The interaction between stressors and habitats is considered when total relative risk calculations are made for each stressor and habitat. In the risk characterization phase the risks for stressors and habitats are compared. Stressors with the greatest potential for ecological impact and habitats most at risk are both identified. This provides a basis for discussions about management of the region.

It is particularly apparent at the regional scale that not all components of the environment can be measured, tested, modeled, or otherwise assessed (Suter 1993a). In addition, there is a large degree of spatial and temporal variability. On a regional scale there is a large degree of uncertainty preliminary risk assessment such as this. However, this should not stop the assessment from proceeding. Uncertainty should be recognized as an inherent component of each stage of the risk assessment and addressed at each stage rather than at the conclusion of the risk analysis. A sensitivity analysis can be performed at the conclusion of the risk analysis to determine how uncertainty is influencing the overall risk rankings.

PROBLEM FORMULATION

The Risk Region

As noted by Suter (1993a), a catchment lends itself to being an easily defined risk region for aquatic-borne contaminants. The catchment considered in this assessment is the Mountain River catchment in southern Tasmania, Australia. It covers approximately 190 km^2 and is located in the Huon Valley region (Figure 8.1).

The Huon Valley is a major horticultural region. The main horticultural crops are apples, cherries, stone fruit, and berries. Apples are by far the biggest crop, and 65% of the Tasmanian apple crop is grown in the Valley, with an estimated market value of $28 million. Other primary industry enterprises include beef cattle production, mushroom farming, herbs, honey, and cut flowers. The Huon Valley is a popular residential locality for urban commuters who have no financial dependence on the land, but value the aesthetic and lifestyle benefits of living in a rural environment.

There is a significant level of public interest and concern in the Huon Valley about environmental issues generally, and waterways in particular. Catchment management in the Huon Valley was formally instigated with the establishment of the Huon Healthy Rivers Project initiated in 1995 with funding provided through federal and local governments. The Huon Healthy Rivers Project is an ongoing project that aims to promote environmental awareness and provide a resource base for community projects.

Information in this assessment was obtained from a number of sources, particularly publications produced by the Huon Healthy Rivers Project and personal communication with Huon Healthy Rivers Project officers who facilitated various

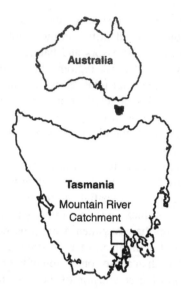

Figure 8.1   Location of Mountain River catchment in southern Tasmania, Australia.

community forums. There has been no extensive or consistent environmental monitoring of freshwater bodies within the valley other than basic water quality data available through state agencies.

Defining Assessment Endpoints within the Mountain River Catchment

Assessment endpoints represent the social values to be protected and serve as a point of reference for the risk assessment. The values to be protected in a region may be described in terms of characteristics of its component populations and ecosystems or in terms of characteristics of the region as a whole (Suter 1993a).

The goals of the local community were used as a starting point for developing assessment endpoints. A community forum, held in 1998 to identify water values for Mountain River as a starting point for setting environmental flows for the river, identified the following issues to be important: improve water quality (particularly decreased *E. coli* counts), maintain/establish water of drinkable and irrigable quality, maintain habitats for aquatic animals, maintain water in suitable volumes to sustain agriculture, maintain catchment quality for town water supply, maintain water for swimming, maintain water for trout fishing, maintain or improve beauty of the river, and maintain seasonal fluctuations between summer and winter flows of the river.

In a 1999 catchment community forum, local residents created an image of their prefered catchment having the following characteristics: clean water that is safe for drinking and swimming, sustainable landuse practices, optimum stream flow, natural vegetation along the riverbanks, an active and responsible community, and an attractive setting for picnics. As noted by Steel et al. (1994), analysis of survey data should consider relationships between survey responses and stakeholder backgrounds. Length and location of residence, occupation, education, and other factors can influence stakeholder values. This particular "community" forum was not well attended by local farmers, and the values stated may not necessarily represent priorities for primary producers. It is vital that assessment endpoints be determined with a conscientious and intelligent effort to represent the values of the entire community.

Beginning with the water body and expanding across the catchment, assessment endpoints were identified based on the views expressed by stakeholders, discussion with resource managers, and expert judgment. The assessment endpoints were identified as:

- Water quality parameters to meet or exceed Australian and New Zealand Guidelines for Fresh Water Quality
- Maintenance of local fish populations (criteria are currently being established by the regulatory body, Tasmanian Inland Fisheries)
- Maintenance of adequate environmental stream flow (criteria are currently being established by the regulatory body, Department of Primary Industries, Water and Environment)
- Maintain or increase native streambank vegetation and reduction of weed density to less than 10% ground cover
- Maintain productive primary industries
- Maintain landscape aesthetics and a good residential environment

Suter (1990) states that good assessment endpoints should have the following characteristics: social relevance, biological relevance (function of its implications for the next higher level of biological organization), unambiguous operational definition, accessibility to prediction and measurement, and susceptibility to the hazard. We compared the above assessment endpoints to Suter's criteria. Water quality parameters to meet or exceed Australian and New Zealand Guidelines for Fresh Water Quality (2000) are currently the only assessment endpoint that meets all of Suter's criteria. At the time of writing the state fisheries agency was in the process of establishing quantitative goals for Tasmanian brown trout fisheries, which is the state's most popular inland fishery, and for establishing an environmental flow for Mountain River. Quantitative goals have currently only been set for the recovery plans of the rare and endangered native galaxias (Crook and Sanger 1997), none of which occur in Mountain River.

The assessment endpoints of maintenance of productive primary industries and landscape aesthetics are intuitively understood, but not well defined. These endpoints do not meet Suter's criteria, but clearly an imperfect definition must not exclude them; maintenance of productive primary industries is of utmost importance in a primarily agricultural catchment. For the purposes of this preliminary relative risk ranking, this assessment endpoint is not operationally defined; instead, general knowledge of good soil and water management practices is applied to it. Work by Landis and McLaughlin (2000) is providing a conceptual framework for quantifying sustainability, although it is unlikely that an unambiguous operational definition for quantifying sustainable agriculture will be achieved because of the huge diversity of inputs to agriculture. It is possible, however, to quantify the sustainability parameters of individual inputs to agriculture, for example, using water quality criteria and regional soil databases.

Similarly, landscape aesthetics is not operationally defined, but can be understood as meaning that Mountain River is a nice place to live. Other assessment endpoints directly impinge on this, particularly the quality of the natural environment as measured through water quality, water flow, weed infestation, aquatic life, and factors affecting agriculture such as soil stability and climate.

Identifying Stressors in the Region

The issues of environmental concern identified in the Huon Healthy Rivers Project were categorized in terms of the stressor and corresponding ecosystem response variable (Table 8.1). With the exception of seasonal flooding, all the stressors identified were anthropogenic. The effects of seasonal floods can be enhanced or mitigated by regional land management practices.

Out of all the stressors identified for the Huon catchment in Table 8.1, the only stressors considered relevant in the risk assessment for Mountain River catchment were agriculture and land clearance for rural residential development. No large-scale forestry activities occur within the catchment, although it is possible there may be some paddock-scale tree plantations on individual farms. No aquaculture occurs within the catchment. Mountain River is too small for boating, and recreational pursuits in the catchment are mainly hiking, horse racing, fishing, and swimming, which were considered to have negligible impact.

Agricultural stressors in the Mountain River catchment were identified as pesticides used in orchards, fertilizers (pasture, orchards, and other cropping activities), pumping irrigation water from the river, weed infestation, and clearing of native bush for farmland (Table 8.1). Another stressor that could be included under the umbrella heading of agriculture is contaminated sites because of possible copper, lead, and arsenic residues in the soil from previous use of orchard pesticides containing these elements. It was decided to omit contaminated sites from this risk assessment because the focus is on risks associated with current agricultural practices. In addition, introducing contaminated sites into the risk assessment involves considerable uncertainty. Currently the actual extent of contamination, if any, is unknown. An intensive regional soil testing program is required before contaminated sites should be considered as a stressor.

Stressors resulting from land clearing for rural residential development were identified as bacteria from septic tank effluent, clearing of native bush for residential purposes, nutrients from households, pumping water from the river for garden and household use, and weed infestation.

Identifying Habitats in the Region

Human exclusion from ecosystems has been symbolic of a long-held belief that somewhere there exists a reference, pristine ecosystem. It is more realistic to recognize that humans are participants in most ecosystems; indeed agricultural ecosystems are created and maintained by humans. It was decided in this risk assessment to recognize anthropogenic habitats in the same way as natural habitats. This has recently been considered as a valid risk assessment approach because changes in

Table 8.1 Anthropogenic Stressors and Ecosystem Response Variables Identified in the Huon Valley

Anthropogenic Stressor	Ecosystem Response Variable
LAND CLEARANCE AND RURAL SUBDIVISION	Species and habitat destruction
	Soil erosion and landslips
	Increase in frequency of erosive flood events
	Increase in environmental weeds — willows, blackberries, ragwort, gorse, pampas grass
INTENSIVE AGRICULTURE Fertilizers and animal waste	Agricultural runoff causing eutrophication of freshwater bodies
	Toxic algal blooms in the estuary affecting estuarine species
Pesticide contamination of soils and water through spray drift, spillage and runoff	Mortality, immunological and reproductive health of local species
	Contaminated sites — it is possible that the lead, copper, and arsenic sprays used earlier last century may have left residues in the soils in older orcharding areas
Soil and water management	Soil erosion and landslips
	Soil compaction and reduction in biological diversity of the soil
	Irrigation water pumped from local waterways, reducing stream flow and changing hydrology, affected microhabitat of aquatic species
RURAL AND COASTAL AREA DEVELOPMENT	River and coast modification altering the habitat of local species
	Wetland degradation; reduction of the "biological filtering" capacity of the estuary
	Septic tank effluent — effluents from improperly maintained septic tanks have contaminated waterways and groundwater in various locations
	Refuse disposal site leachate — current public sites are located at Huonville, Geeveston, Cygnet; former sites were located at Glen Huon and Judbury; older and former public and private sites are spread throughout the municipal area; contaminants of unknown types and quantities discharged to waterways
	Pumping drinking and household water from local waterways, reducing stream flow and changing hydrology affecting the microhabitat of aquatic species
	Nutrient input from sewage; sewage treatment plants are located at Ranelagh, Cygnet, and Geeveston; sewage lagoons at Huonville, Dover, Southport; Franklin sewage currently discharged into the Huon River
	Solid waste management — public landfill facilities at Geeveston and Cygnet; waste transfer stations at Cygnet, Southport, Dover, and Huonville; private contractor also provides recycling facilities at each site
	Untreated stormwater containing unknown types and quantities of contaminants
FORESTRY	Soil erosion and landslips
	Nutrient runoff
	Road building causing siltation of waterways
	Environmental weeds
AQUACULTURE	Nutrients from fish waste, uneaten food, and disposal of net wash effluent causing nutrient enrichment of the estuary and increasing probability of toxic algal blooms
	Escaping fish possibly competing with native species
RECREATIONAL PURSUITS IN THE VALLEY	Ballast water introducing pest species
	Boat pollution (fuel, sewage waste, rubbish)

Information from the Huon Healthy Rivers Project (1997) was used as a basis for this table.

Table 8.2   Habitats Identified within the Mountain River Catchment

Habitat	Description	Major Impacts Within
Aquatic	All water bodies are included in this category, although the emphasis is on larger waterways in the catchment, in particular Mountain River and Crabtree Rivulet	Contamination of the water body Eutrophication Changes in hydrology
Native vegetation	Includes all native vegetation types mapped in the TasVeg™ 2000 series; priority vegetation associations in the Mountain River catchment are *Eucalyptus ovata, E. amygdalina, E. tenuiramis, E. globulus*	Reduction in the habitat of native flora and fauna
Orchard	Includes all land mapped as orchard; major orchard crops are apples, followed by cherries	Reductions in populations of beneficial insects Weeds competing with orchard trees, especially during establishment
Pasture	Includes all pastures used for grazing sheep, horses, goats, and for cutting hay; limited crop production in Mountain River catchment, but any occasional cropping that does occur is also included in this category	Weeds competing with pasture and crop species; weeds can also decrease quality of pasture and decrease price of cut hay
Residential	Includes the area around each residence that is actively used or maintained by the resident; also includes the residence	Loss of aesthetic value

ecological systems result in risks to cultural resources, economic activity, and quality of life because of the numerous and important services of nature (Suter 1999a). Moreover, ecological risks can often be considered as risks to the sustainability of the activities being assessed (Suter 1999b).

Based on landuse in the catchment, five different habitat categories were identified (Table 8.2). Given the diversity of stressors, there are a variety of impacts that could occur within each habitat.

Interaction of Stressors and Habitats — Risk Hypotheses in the Conceptual Model

At this point in our preliminary risk assessment, stressors and habitats in the region have been identified. The values of various stakeholder groups have been considered in the formulation of assessment endpoints. A conceptual model of the region showing the interaction of stressors, habitats, and the potential for impacts on chosen assessment endpoints is given in Figure 8.2. The conceptual model describes the approach that will be used for the risk analysis phase. It is a graphical summary of the risk hypotheses being assessed within the catchment (USEPA 1992). Conceptual models are representations of the assumed relationships between sources and effects (Suter 1999a). The conceptual model shown in Figure 8.2 represents assumed interactions of stressors and habitats within the catchment. It contains uncertainty; however, it is adopted as an operating tool in the absence of more complete knowledge.

The risk hypotheses shown in Figure 8.2 assume that agriculture and land clearing for rural residential areas produce multiple stressors that have potential for

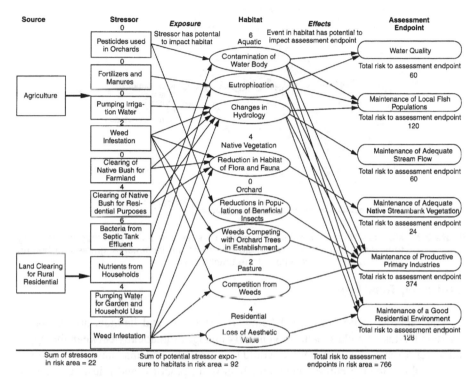

Figure 8.2 Conceptual model — hypothesized interactions between stressors and habitats in Mountain River. The rankings and calculations shown here are for risk region 4. (Figure drawn by Angela Schuler.)

contamination of local water bodies, eutrophication, changes in hydrology, reduction in the habitat of native flora and fauna, reductions in populations of beneficial insects, increased weed competition in pastures, and loss of aesthetic value in residential areas.

Particular emphasis has been placed in this regional risk assessment on the conceptual model as a tool for visually interpreting the relative risk calculations. This is described in the Risk Analysis section.

RISK ANALYSIS USING THE RELATIVE RISK MODEL

Much of the input data for the risk analysis in this assessment came from landuse patterns shown in the Tasmania 1:25,000 Series. The map sheets used were Longley 5024 (Edition 2, 1988) and Huonville 5023 (Edition 2, 1987). Digitized map data are supplied to the Australian public on a cost recovery basis and there is only a limited amount of digitized data available. Landuse themes, including vegetation, were not available in digital format so it was necessary to digitize vegetation patterns from paper maps. The maps were scanned and on-screen digitized. Vegetation themes were transformed from scan unit coordinates to the Universal Transverse Mercator projection using Shape Warp 2.2. ArcView® version 3.1 (Environmental Systems Research Institute, Redlands, CA) was the GIS software used in this assessment.

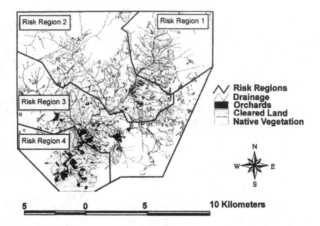

Figure 8.3   Mountain River catchment risk regions. Risk regions have been mapped according to flow of tributaries into Mountain River and the incremental gradient of human activity in the lower reaches of the catchment.

Identifying Risk Areas

The ranking criteria described below for stressors and habitats are primarily based on landuse. Landuse patterns generally change dramatically between the upper and lower reaches of a catchment. It would be unrealistic and unachievable to attempt a risk ranking for the entire catchment. It is more practical and relevant to divide the region into subareas or risk regions so that stressors and habitats within a specific subarea can be better considered. This also allows comparison of risks from different stressors to specific habitats within different catchment areas.

An incremental gradient of human activity occurs as Mountain River flows down through the catchment. The intensity of agriculture, orcharding, and residential development increases. The risk regions in Figure 8.3 were chosen to match this gradient of human activity and of the natural boundaries determined by contours and tributaries flowing into Mountain River. Aligning risk regions with the flow of tributaries to Mountain River was very important. Even though two tributaries may at some point only be separated by a few kilometers, they may flow through very different landuse activities before they join the main channel, ultimately contributing very different inputs to the main channel.

Ranking Stressors

The most accessible data about the Mountain River catchment came from the 1:25,000 map series. However, it was not possible to quantify the extent and severity of each stressor by simply studying landuse maps. Some quantitative data were available for the Mountain River catchment, although the lack of coordination between government agencies made it difficult to access. Surprising gaps in the knowledge about potential environmental stressors were discovered. The local council did not have a database that could identify how many people lived within the physical catchment nor how many septic tanks were installed within the catchment,

Table 8.3   Ranking Criteria for Mountain River Habitats

Habitat	Rank Criteria and Assigned Points	Uncertainty
Aquatic	6; The aquatic habitat was given a single, high ranking because all activities that occur within a catchment ultimately impinge on the waterway. In addition, the risk regions all include a section of the waterway, so the aquatic habitat was considered to be a highly ranked habitat in all risk areas	Assumption that a high ranking is justified across all regions
Native Vegetation	6; 23–37% (of total catchment native vegetation found within the risk region) 4; 11–22% 2; 10% 0; < 1%	Accuracy of 1:25,000 maps
Orchard	6; 41–46% (of total catchment orchards found within the risk region) 4; 15–40% 2; 1–14% 0; <1%	Accuracy of 1:25,000 maps
Pasture	6; 29–35% (of total catchment pastures found within the risk region) 4; 16–28% 2; 1–15% 0; < 1%	Accuracy of 1:25,000 maps Classification of pasture vs. vacant land and home gardens
Residential	6; Many ratepayers (approx > 5000) 4; Not so many ratepayers (approx > 3000 2; Few ratepayers (approx > 1000) 0; No ratepayers	No localized population data available for different areas of the catchment; assumption is that all residences are contributing equally to the source

although they could make approximations. A total of 39 household water pumps are installed along the river, although locations of the pumps could only be estimated based on residential density along the river.

To collect quantitative data about the extent and severity of each stressor in each risk region is a task not warranted for a preliminary risk assessment of the catchment. All data currently available about the catchment were collated, but no additional field data were collected for this preliminary risk assessment.

We decided in this preliminary risk assessment to use expert knowledge of the region and landuse maps to qualitatively rank stressors. The ranking criteria and points assigned were:

6   Likely to occur
4   Possibly could occur
2   Unlikely to occur
0   Very unlikely to occur

The distance between risk categories is assumed equal, i.e., stressors ranked as 6 are not by definition three times larger than those ranked as 2. This also applies to the habitat ranking criteria given in Table 8.3.

Uncertainty is obviously a significant consideration at this point of the risk assessment. However, it was planned to undertake a sensitivity analysis to determine if the stressor rankings had a significant effect on the relative risk ranks. In this way, the effect of an incorrect stressor ranking on the overall relative risk outcomes could be compared. The sensitivity analysis is described in the Risk Characterization section.

The ranking criteria and points assigned for severity of weed infestation were based on standard categories for mapping weed density as used in the survey conducted by the Huon Healthy Rivers Project.

6   Heavy > 50% ground cover
4   Moderate 25 to 50% ground cover
2   Light weeds 10 to 25% ground cover
0   Scattered weeds < 10% ground cover

Ranking Habitats

Habitats were ranked according to the proportion of a particular habitat within a region. To determine the proportion of a particular habitat within a risk region, map themes were manipulated and planimetric areas measured using ArcView software. Habitat ranks and uncertainties associated with the ranking are described in Table 8.3.

The major source of uncertainty in establishing the ranking criteria for habitats stems from the 1:25,000 maps. The content of these maps was determined from aerial photography undertaken in 1986. Obviously, there would have been changes in landuse since that time, so the exact proportion of different vegetation and landuse types would have changed. However, in the absence of other map data, the 1:25,000 series maps must be used. Since 1986, the major changes in landuse in the catchment have been the subdivision of pasture into rural residential blocks. The number of residences in the catchment is now greater than indicated on the maps and, consequently, the true extent of pasture may have been overestimated. However, there have not been other significant landuse changes in the catchment, and for the purposes of this preliminary risk assessment the 1:25,000 maps were considered adequate.

Relative Risk Calculations Using the Conceptual Model

Figure 8.2 is the conceptual model for this risk assessment. Visually it describes all the interactions between stressors and habitats being considered in this risk assessment. It has been produced as a spreadsheet so that it can simultaneously mathematically describe the risks associated with the stressors and habitats found in each risk region, based on the assumed interactions between stressors and habitats. These assumed interactions are indicated by the exposure and effects arrows.

There is a number above each stressor and habitat category in Figure 8.2. These numbers are the risk rankings for the stressor and habitat in a given risk region (the example reproduced here is for risk region 4). This number describing risk ranking is in a cell that is part of a spreadsheet formula. Spreadsheet formulas are used to calculate the risks indicated in Figure 8.2, that is, the sum of stressors within the risk region, sum of potential stressor exposure within the risk region, total risk to assessment endpoints within the risk region, and total risk to each assessment endpoint. Incorporating spreadsheet calculations into the conceptual model means it is easy to compare total risks between different risk regions and for different rankings of stressors and habitats. The assumed interactions between stressor and habitat remain constant; only the risk rankings change.

The spreadsheet formulas used for calculating risk are:

- Sum of stressors in risk region = Σstressors
- Sum of potential stressor exposure in risk region = Σ(stressor × habitat) for interactions where an exposure arrow indicates the stressor has potential to impact habitat
- Total risk to assessment endpoint = Σ(stressor × habitat) for interactions where an exposure arrow indicates the stressor has potential to impact habitat AND an effects arrow indicates that an event in the habitat has potential to impact assessment endpoint
- Total risk to assessment endpoints in risk region = Σ(total risk to assessment endpoint)

The use of exposure and effects arrows serves to ensure that only realistic interactions are included in the conceptual model and the risk ranking. Not every stressor has the potential to impact every habitat, nor has every stressor the potential to impact every assessment endpoint. A relative risk ranking cannot simply be a sum of stressor × habitat; the interactions assumed in the conceptual model must be accounted for. These interactions are indicated by the linking arrows in Figure 8.2.

When Wiegers et al. (1998) did their relative risk calculations for the Port Valdez regional risk assessment, they used an exposure filter and effects filter to ensure that only realistic interactions were included in the risk calculations. Their filtering method included 1 in the risk calculations that represented realistic interactions and 0 in the interactions that represented unrealistic interactions. Their method did not use the conceptual model as visual reference, so the filtering method involved individually assessing each stressor/habitat/impact interaction and questioning whether it was a realistic scenario.

RISK CHARACTERIZATION

A comparison of risks to assessment endpoints in risk region 4 is shown in Figure 8.4. This is where the most agricultural and residential development in the catchment has occurred. Risks to productive primary industries are greatest, which is not surprising considering the diversity of inputs to agriculture. After primary industries, risks to the residential environment and maintenance of fish populations are greatest. Degradation of water quality in the region had the greatest impact on assessment endpoints. Initially it was surprising that risks to native vegetation were comparatively low, given the development that has occurred in the region. However, this risk outcome is accurate because there is actually very little native vegetation remaining in the region (habitat rank is 2), so risks to this habitat type are relatively low.

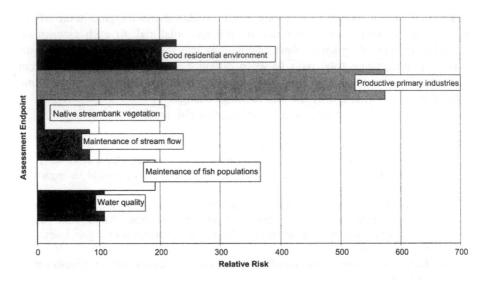

Figure 8.4   Comparison of risks to assessment endpoints in risk region 4. Relative risks rather than absolute values describe environmental conditions in the catchment. (See color insert following page 178.)

A comparison of risks to assessment endpoints in different risk regions is shown in Figure 8.5. The same general trends appear throughout the catchment, with risks greatest to productive primary industries, followed by residential environment and maintenance of fish populations. Generally, risks to all assessment endpoints are greater in risk regions 3 and 4 because more agricultural and residential development has occurred there.

It is important when interpreting the risk outcomes to remember that *relative* risks form the basis of the RRM. A risk outcome in itself has no meaning unless compared to other risk outcomes — in this way risks are prioritized. One limitation of the RRM is that stressors and habitats are ranked on relative likelihood of occurrence, not on relative consequence of occurrence. Different stressors can have different effects with different consequences for the habitat of concern. Ranking of stressors on the basis of consequence of occurrence was not attempted in this preliminary study due to a limited understanding of ecological processes within the catchment. With limited scientific data available, rankings of ecological consequence would tend to be value driven, rather than factual.

Sensitivity Analysis

As previously mentioned, the ranking of stressors is a major source of uncertainty in the assessment. Sensitivity analysis is a means to assess the robustness of the model — how would the risk outcomes change if we had new knowledge that allowed more precise ranking of the stressors?

Weed infestation is a significant stressor in the catchment, particularly as it has been included as a stressor to agricultural and residential environments. The initial rankings of weed infestation came from a 1999 Huon Healthy Rivers weeds survey.

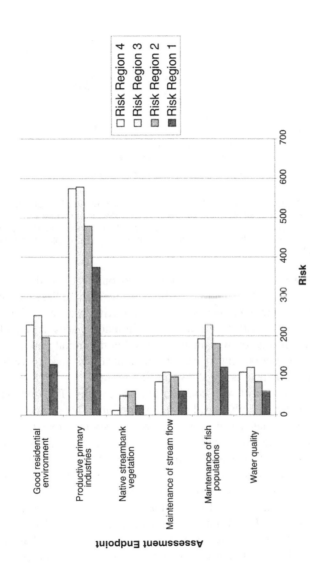

Figure 8.5 Comparison of risks to assessment endpoints between risk regions. The increased risks to assessment endpoints in risk regions 3 and 4 reflect the increased level of human activity in this part of the catchment. (See color insert following page 178.)

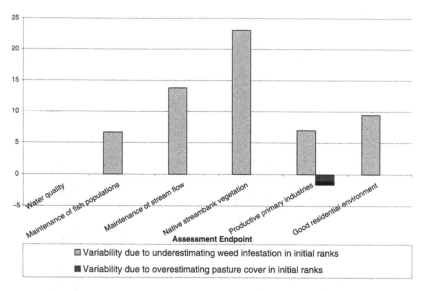

Figure 8.6   Sensitivity analysis: the effect that changing initial rankings has on the risk outcomes. The conceptual model is sensitive to changes in ranking of weed infestation, but insensitive to changes in ranking of pasture cover. Uncertainty in some
parameters is more critical than uncertainty in others. (See color insert following
page 178.)

However, this survey was preliminary. Weed cover was only estimated along a few
roads in each risk region and only certain weed species were included in the survey.
No survey was conducted along the banks of Mountain River where crack willows,
blackberries, hawthorn, and thistles have taken over in many places. It is likely that
the true extent of weed infestation was underestimated in the initial rankings. Weed
infestation was given a higher ranking in each risk region, and the risks to assessment
endpoints recalculated. The results are shown in Figure 8.6.

Weed infestation is a stressor with considerable uncertainty. The true extent of
pasture cover is a habitat type with considerable uncertainty. This stems from using
1987 1:25,000 maps to calculate pasture cover. As previously mentioned, since 1987
various farms have been subdivided and the pasture converted to rural residential
blocks. It is possible that the calculated land areas overestimate the true extent of
pasture cover. Pasture cover was given a lower ranking in each risk region, and the
risks to assessment endpoints recalculated (Figure 8.6).

Uncertainty in the ranking of weed infestation (stressor) can have considerable
effect on the risk outcomes (Figure 8.6). If the true extent of weed infestation is
actually greater than initially estimated, then the actual risks to several assessment
endpoints are greater. The risks to native streambank vegetation increased by 23%,
and the risks to maintenance of stream flow increased by nearly 14%. This occurs
because more weeds are competing with native vegetation and more willows are
interfering with natural watercourses.

Figure 8.6 also shows that uncertainty in the ranking of pasture cover (habitat)
has negligible effect on the risk outcomes. If the true extent of pasture cover is
actually less than initially estimated, the actual risks to assessment endpoints remain

unchanged except for a 2% decrease in risk to productive primary industries. This is explained by the fact that pasture is an input to agriculture, and if there is less pasture cover there is less risk to productive agriculture. Obviously this does not account for competition between agricultural and residential landuses. If there is less pasture cover, it may be that agriculture is actually at more risk because there is less viable land for farming.

A Basis for Action

The sensitivity analysis shows that the conceptual model is more sensitive to uncertainties in some parameters than in others. An on-paper exercise such as this can be used to highlight the most important knowledge gaps about the catchment. In the example seen in Figure 8.6, obviously it is more important to invest time and money into a comprehensive weeds survey rather than precisely determine the percentage cover of each landuse type.

A preliminary study such as this can also provide a basis for establishment of environmental monitoring programs. Monitoring programs should be concentrated in risk regions 3 and 4, as this is where most environmental stressors are impacting. Maintenance of productive primary industries is most at risk, so monitoring should focus on the stressors that are having an impact on this assessment endpoint. In the conceptual model (Figure 8.2) it is clear that changes in the aquatic habitat (water contamination, eutrophication, and hydrology) have a major impact on this assessment endpoint. There is a priority to monitor the stressors that are impacting the aquatic environment. Nutrient levels, bacterial counts, and pesticide residues should be included in routine water quality monitoring programs.

The risk analysis can also provide information for local environmental groups who undertake activities in the catchment. Figure 8.2 reinforces that weed control is one of the most effective means for the community to achieve its environmental goals.

DISCUSSION: REGIONAL RISK ASSESSMENT

Because there is an increasing need to incorporate multiple stressors into all ecological risk assessments, the RRM provides a methodology that is well suited to regional applications. Our preliminary risk assessment has demonstrated the utility of the RRM methodology as a tool for assessing the risks associated with agriculture and rural development in a small catchment.

The RRM is a straightforward approach to risk assessment — more than anything it is a framework for data collection and decision making. In a preliminary risk assessment such as this, perhaps the most important function is collation of information about the region, with a focus on what stakeholders want for the region. Identification of assessment endpoints represents a crucial but difficult part of this process. Currently, practical regional risk assessments are somewhat limited by the generalist goals that are chosen in the absence of more specific baseline data. Minimal scientific and regulatory information was available for Mountain River, and this greatly hindered the definition of concise assessment endpoints.

The conceptual model formed the basis of the entire risk assessment. The risk outcomes were based entirely on the interactions between stressors and habitats assumed in this model. This particular conceptual model is unique to the Mountain River catchment. The conceptual model for another catchment would be different based on the interactions that occur in that environment. For example, soil erosion and salinity caused by tree removal are not considered to be environmental problems in the Mountain River catchment, but are significant issues in many other Australian catchments; therefore, their inclusion would require development of an entirely different conceptual model.

The model was formulated based on all available knowledge about the region. However, it is possible that different risk assessors with similar knowledge of the region might propose a different conceptual model to form the basis of the risk assessment. The risk outcomes might be different, but the framework provided by the RRM means there is a tangible basis for discussion of regional environmental priorities.

The RRM framework provided an effective utilization of time and resources for gaining an understanding of ecological issues within the Mountain River catchment. Preliminary risk assessments such as this are valuable tools in planning expensive fieldwork and ground truthing projects. The mapping component of the RRM approach is particularly useful for identifying study areas with the highest incidence of environmental stressors and to focus fieldwork on environmental hot spots. With the advent of paddock mapping for a number of Tasmanian cropping industries (e.g., pyrethrum and poppies), it will soon be possible to target catchments with particular cropping activities. Statewide, there is increasing interest in estimating environmental impacts of large-scale plantation forestry (Future Challenges of Agriculture Public Forum, Devonport, Tasmania, 2003).

One property of the RRM that is potentially misleading is the system for ranking of habitats. Habitats are ranked based on their occurrence within the region, so that a small area of one habitat is given a lower ranking than a large area of another. This system is flawed when endangered habitats form part of the region. For a rare habitat, it is not logical to conclude that because there is less land area of that habitat, the risks to it are less.

ACKNOWLEDGMENTS

We express our appreciation to Eugene Hoerauf for his GIS assistance. The International Exchanges Program between the University of Tasmania and Western Washington University, Bellingham, Washington, facilitated this research.

REFERENCES

Australian and New Zealand Guidelines for Marine and Fresh Water Quality. 2000. Australian and New Zealand Environment Conservation Council, Canberra. In revision.

Crook, D. and Sanger, A. 1997. Recovery Plan for the Peddar, Swan, Clarence, Swamp and Saddled Galaxias, Inland Fisheries Commission Report, Hobart, Australia.

Environmental Systems Research Institute, Inc. 1998. ArcView® Version 3.1. Redlands, CA.

Hunsaker, C.T., Graham, R.L., Suter, G.W., II, O'Neill, R.V., Jackon, B.L., and Barnthouse, L.W. 1989. Regional Ecological Risk Assessment: Theory and Demonstration, ORNL/TM-11128, Oak Ridge National Laboratory, Oak Ridge, Tennessee.

Huon Healthy Rivers Project. 1997. Huon Valley Catchment Management Plan. Huon Valley Council, Tasmania.

Landis, W.G. and Wiegers, J.A. 1997. Design considerations and a suggested approach for regional and comparative ecological risk assessment, *Hum. Ecol. Risk Assess.*, 3, 287–297.

Landis, W.G. and McLaughlin, J.F. 2000. Design criteria and derivation of indicators for ecological position, direction and risk, *Environ. Toxicol. Chem.*, 19, 1059–1065.

Landis, W.G., Luxon, M., and Bodensteiner, L.R. (in press). Design of a relative rank method regional-scale risk assessment with confirmatory sampling for the Willamette and McKenzie Rivers, Oregon, in *Ninth Symposium on Environmental Toxicology and Risk Assessment: Recent Achievements in Environmental Fate and Transport*, Price, F.T., Brix, K.V., and Lane, N.K., Eds., ASTM STP1381, American Society for Testing and Materials, West Conshohocken, PA.

Landis, W.G., Matthews, G.B., Matthews. R.A., and Sergeant, A. 1994. Application of multivariate techniques to endpoint determination, selection and evaluation in ecological risk assessment, *Environ. Toxicol. Chem.*, 13, 1917–1927.

NRC (National Research Council). 1983. *Risk Assessment in the Federal Government: Managing the Process*. National Academy Press, Washington, D.C.

Steel, B.S., List, P., and Shindler, B. 1994. Conflicting values about Federal forests: a comparison of national and Oregon Publics, *Soc. Natur. Resour.*, 7, 137–153.

Suter, G.W., II. 1990. Endpoints for regional ecological risk assessments, *Environ. Manage.*, 14(1), 9–23.

Suter, G.W., II. 1993a. *Ecological Risk Assessment*, Lewis Publishers, Chelsea, MI.

Suter, G.W., II. 1993b. A critique of ecosystem health concepts and indexes, *Environ. Toxicol. Chem.*, 12, 1533–1539.

Suter, G. W., II. 1999a. Developing conceptual models for complex ecological risk assessments, *Hum. Ecol. Risk Assess.*, 5, 375–396.

Suter, G.W., II. 1999b. A framework for assessment of ecological risks from multiple activities. *Hum. Ecol. Risk Assess.*, 5, 397–413.

USEPA. 1992. A Framework for Ecological Risk Assessment, EPA/630/R-92/001, Risk Assessment Forum, Washington, D.C.

Wiegers, J.K., Feder, H.M., Mortensen, L.S., Shaw, D.G., Wilson, V.J., and Landis, W.G. 1998. A regional multiple-stressor rank based ecological risk assessment for the Fjord of Port Valdez, Alaska, *Hum. Ecol. Risk Assess.*, 4, 1125–1173.

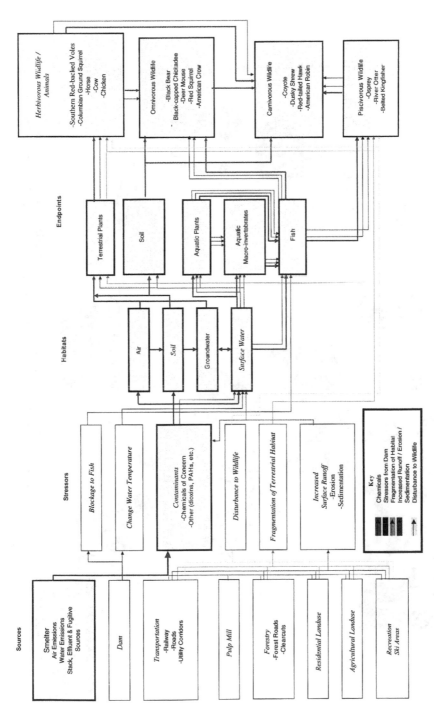

Figure 2.4 Example of a conceptual model incorporating the basic framework of the relative risk model (designed by Hart Hayes).

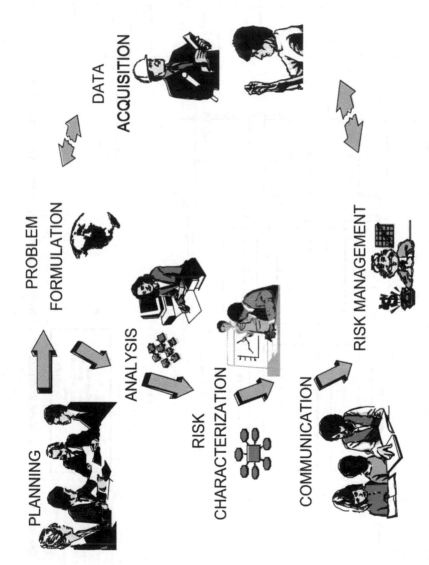

Figure 3.1 The importance of risk management as well as how problem formulation feeds into data acquisition and data analysis. (From Cirone, P.A. and Duncan, P.B., *J. Hazardous Mater.*, 78, 1–17, 2000. With permission.)

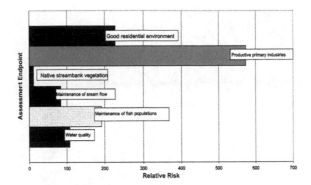

Figure 8.4 Comparison of risks to assessment endpoints in risk region 4. Relative risks rather than absolute values describe environmental conditions in the catchment.

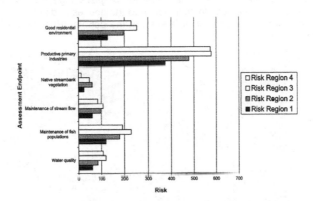

Figure 8.5 Comparison of risks to assessment endpoints between risk regions. The increased risks to assessment endpoints in risk regions 3 and 4 reflect the increased level of human activity in this part of the catchment.

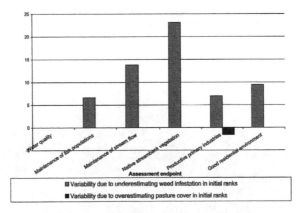

Figure 8.6 Sensitivity analysis: the effect that changing initial rankings has on the risk outcomes. The conceptual model is sensitive to changes in ranking of weed infestation, but insensitive to changes in ranking of pasture cover. Uncertainty in some parameters is more critical than uncertainty in others.

Figure 9.1   Aerial photograph of *Bairro da Serra,* the largest village in the vicinity of PETAR. Clear cuttings are seen in conjunction to the houses and in the upper parts of the photograph. The Betari River is seen at the bottom. (Photograph Courtesy of H. Shimada.)

Figure 9.2   Piles of waste rock near an abandoned lead mine located in the vicinity of PETAR. (Photograph courtesy of R. Moraes.)

Figure 9.3   A quarry for limestone mining located in the vicinity of PETAR. (Photograph courtesy of H. Shimada.)

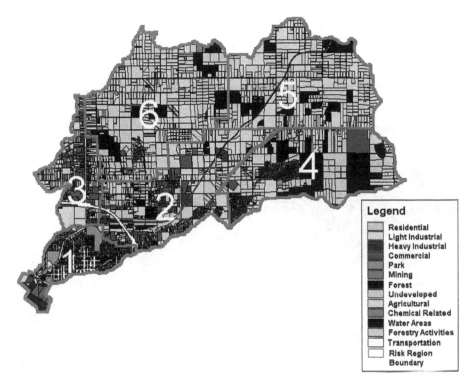

Figure 10.2   Risk regions and landuses in the study area.

Legend
Residential
Light Industrial
Heavy Industrial
Commercial
Park
Mining
Forest
Undeveloped
Agricultural
Chemical Related
Water Areas
Forestry Activities
Transportation
Risk Region
Boundary

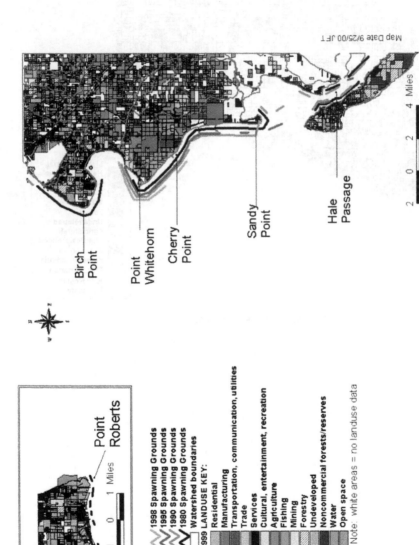

Figure 11.1 CP region herring spawning grounds: 1980, 1990, 1995, and 1998. (Prepared by J. Thomas, data from Gonyea et al. 1982; Stick, 1990, Lemberg et al. 1997; Whatcom County Planning and Development Services, 2001.)

Map Date 9/25/00 JFT

Birch
Point

Point
Whitehorn

Cherry
Point

Sandy
Point

Hale
Passage

2   0   2   4 Miles

Point
Roberts

0   1 Miles

1998 Spawning Grounds
1995 Spawning Grounds
1990 Spawning Grounds
1980 Spawning Grounds
Watershed boundaries
1999 LANDUSE KEY:
Residential
Manufacturing
Transportation, communication, utilities
Trade
Services
Cultural, entertainment, recreation
Agriculture
Fishing
Mining
Forestry
Undeveloped
Noncommercial forests/reserves
Water
Open space
Note: white areas = no landuse data

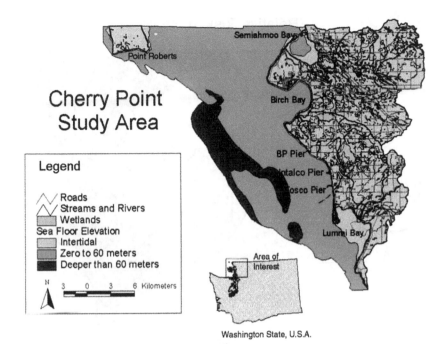

Figure 13.1 The Cherry Point study area in Northern Whatcom County, Washington. BP (British Petroleum) Oil Company, Alcoa Intalco Works Aluminum, and Tosco Oil Company maintain shipping piers on the coast.

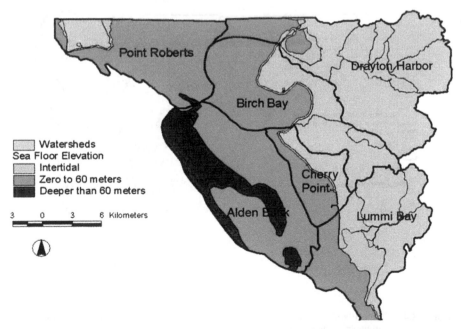

Figure 13.2 The study area divided into six subregions based on watershed and bathymetric boundaries.

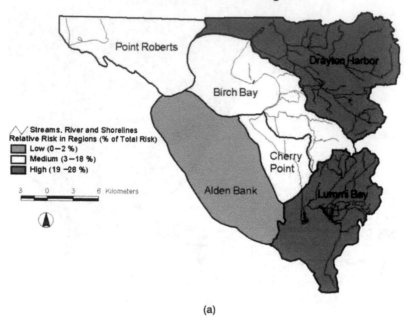

(a)

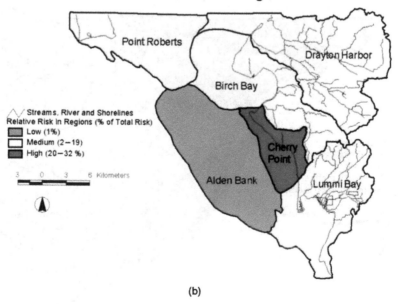

(b)

Figure 13.4 Comparisons of the relative risks depending upon assumptions about the sensitivity of the habitat vs. area.

CHAPTER 9

Establishing Conservation Priorities in a Rain Forest Reserve in Brazil: An Application of the Regional Risk Assessment Method

Rosana Moraes, Wayne G. Landis, and Sverker Molander

CONTENTS

INTRODUCTION

The Parque Estadual Turístico do Alto Ribeira (PETAR) is a natural reserve in southeastern Brazil that was selected for an ecological risk assessment (ERA) due to the importance of its preservation. Its speleological patrimony and its great number of endemic and threatened species motivated UNESCO to declare PETAR as an International Biosphere Reserve and a World Heritage Site (Lino 1992; UNESCO 2001). One of the largest and best preserved areas of Atlantic rain forest in the country is constituted by PETAR and its neighboring reserves: Fazenda Intervales State Park, Serra do Mar Environmental Protection Area, Xitue Ecological Station, and Carlos Botelho State Park. This type of rain forest has been ranked as number four among the so-called "hot spots," i.e., areas featuring an exceptional concentration of endemic species and experiencing an exceptional loss of habitat (Myers 2000). Of the remaining primary vegetation of the forest, which is merely 7% of its original extent, only about a third is protected in parks or reserves. These reserves were primarily established to promote the long-term conservation of biodiversity. This goal should be reached by maintaining natural processes and viable populations, and by excluding human threats. However, most of these reserves do not fulfill their roles because of disturbances, transformation to intensive uses, and fragmentation of habitat due to human settlements within and in the surroundings of the reserves.

Another reason for choosing PETAR as a case study was its social and economic contexts. The park is located in the southwestern part of São Paulo State in the Ribeira Valley, one of the poorest and least developed areas of the state (SMA 1997). Agriculture is the main economic activity in the Ribeira Valley, but its expansion faces limitations related to land occupation and cropping preconditions such as soil, topography, and climate. Despite the fact that PETAR represents one of the most important protected areas of Atlantic rain forest in Brazil, any management action must achieve a balance between the costs of a great limitation of the economic growth of the region (since alternative activities for the population are still very scarce) and the benefits of preservation. The regional risk assessment (RRA) approach was justified by the complexity of the PETAR case: multiple stressors derived from diverse sources widely spread over a large geographical area.

PLANNING PROCESS

The EcoRA, of which the RRA was a part, started with interviewing stakeholders at regulatory agencies, local villages, the PETAR administration, local farms, research institutes, and environmental groups in order to understand public, ecological, and management concerns and to evaluate the status of the available information.

The main concern of the public was that the rivers provide fish and clean water. Another concern was the aesthetic value of the park, which brings tourism income for many residents.

According to regulatory agencies, residents, and researchers, ecological concerns included the preservation of the forest and the protection of caves, which are representative of the region (Sanchez 1984). Conservation actions already taken at governmental levels include specific federal and state acts for special protection of systems of caves and karstic formations, and all remains of the Atlantic rain forest.

From the management point of view, the main concerns were related to human activities, such as mining, palm tree extraction, and the establishment of new settlements, which are threatening the preservation of biota in and near the park. In 1985, a management plan for the park (IF 1985) was designed to deal with some of these environmental problems, but its implementation was discontinued due to a shortage of staff, infrastructure, and funds. The expansion of the park to include headwaters of rivers that flow through the reserve was introduced in that plan as one of the management alternatives for reducing environmental impacts inside PETAR.

The scope of the EcoRA was decided on the basis of these concerns and the evaluation of local and scientific knowledge. The assessment was a retrospective one and included the entire watersheds of the rivers flowing through PETAR. The ecological value to be protected, the assessment endpoint according to USEPA Guidelines for Ecological Risk Assessment (USEPA 1998), was the quality of its rivers, despite the fact that most stressor sources are related to landuse practices. A reason for this choice was that changes anywhere in the landscape are likely to influence rivers and watercourses as collectors and transmitters of the stressors from the sources to downstream areas (Karr 1998). This is especially pertinent in the PETAR region, where rivers are very numerous and their networks include connected surface and subterraneous segments.

PROBLEM FORMULATION

The objective of the RRA was to identify stressors with the greatest potential for ecological impact, habitats most at risk, and subareas inside the watersheds of the main surface and subterraneous rivers flowing through PETAR that are more likely to be impacted. Such information is important when establishing conservation priorities.

There are approximately 15 small villages in and near the park (Figure 9.1) (SMA 1991). Since the majority of the houses do not have sewage treatment systems, domestic wastewater is discharged directly into watercourses. This source of organic material may cause a decrease in oxygen concentrations of the water close to the settlements due to the breakdown of organic compounds, while, at greater distances, the outlets may cause an increase of inorganic nutrients in the water (SEPA 1993). Agricultural activities at small scale, which is rather common in the region, may also contribute to a greater availability of phosphorus in the watercourses by increasing soil erosion after clear-cutting areas of the forest.

Intermediate-scale agriculture also occurs in and near PETAR, where tomatoes and other crops are cultivated. Increased erosion after clear-cutting and during cultivation and use of fertilizers may increase the amount of inorganic nutrients in water (Cullen et al. 2001). Pesticides used on crops, or applied near the villages by health authorities to control vector populations, can reach the aquatic environment of the reserve by surface runoff or leakage through the soil from the site of application. Wind drift of pesticides may also occur, as well as several other possible

Figure 9.1   Aerial photograph of *Bairro da Serra,* the largest village in the vicinity of PETAR. Clear cuttings are seen in conjunction to the houses and in the upper parts of the photograph. The Betari River is seen at the bottom. (Photograph courtesy of H. Shimada.) (See color insert following page 178.)

Figure 9.2   Piles of waste rock near an abandoned lead mine located in the vicinity of PETAR. (Photograph courtesy of R. Moraes.) (See color insert following page 178.)

routes of pesticide exposure. Sensitive aquatic populations may decline if the concentration of pesticides in water increases to levels high enough to cause effects, unless the populations have adapted to contaminated conditions.

Another source of stressors is former mining operations. Mining has been an important activity in Ribeira Valley since the 17th century. Extraction of metals declined during the 1980s due to environmental restrictions and problems with the economy. There are 7 gold and 44 lead open pits or underground abandoned mines in the watersheds of rivers flowing though PETAR (Shimada, Burgi, and Silva 1999). Degraded lands were seldom reclaimed because mining companies went into bankruptcy or were indifferent, a problem made worse by the lack of effective enforcement of existing environmental legislation (Macedo 2000). As a result, piles of waste

Figure 9.3 A quarry for limestone mining located in the vicinity of PETAR. (Photograph courtesy of H. Shimada.) (See color insert following page 178.)

rock and deforested areas are found near abandoned mines (Figure 9.2). Monitoring studies done by the Environmental Sanitary Agency of São Paulo State revealed high concentrations of heavy metals in one of the rivers draining such an area (CETESB 1988; CETESB 1991; Eysink et al. 1988). Chronic exposure to heavy metals may, for instance, influence the structure and composition of aquatic communities by loss of sensitive species (Petersen 1986; Beltman et al. 1999).

Limestone mining is another source of stressors. Four quarries are located in or near PETAR (Cullen et al. 2001) (Figure 9.3). After blasting explosions, a considerable amount of particles are in the air and deposited in surface water, increasing water turbidity. The suspended solids may settle on the bottom over time, which may affect the benthic community. Particles suspended in streams may decrease light penetration and, after settling, reduce pool-riffle differentiation, therefore, affecting both the benthic community and the availability of suitable habitat patches for larger individuals (Bekerman and Rabeni 1987).

Illegal harvesting of palm tree hearts, much appreciated as food, is also a problem in PETAR. This type of palm has become almost extinct because harvesting includes killing the whole tree (Galetti and Fernandez 2002). Although palm tree exploitation is a management problem in PETAR, it was not considered a potential stressor source in this assessment as it is unlikely to affect the quality of its rivers.

The selected sources for the assessment were agriculture, human settlements, and mining for limestone, lead, and gold, all of which release chemical (pesticides, nutrients, and metals) and physical (particles) stressors. Figure 9.4 shows distribution of the selected stressor sources and watershed limits.

The assessment endpoint was the self-sustaining aquatic fauna of PETAR. This selection was mostly driven by societal values, in this case, the ability of the rivers to provide fish and clean water and the long-term maintenance of biodiversity of the rivers. The epigean (surface) and hypogean (subterranean) watercourses represented the habitats of concern. A simple generic conceptual model illustrating the stressor pathways from the sources to the assessment endpoint is presented in Figure 9.4.

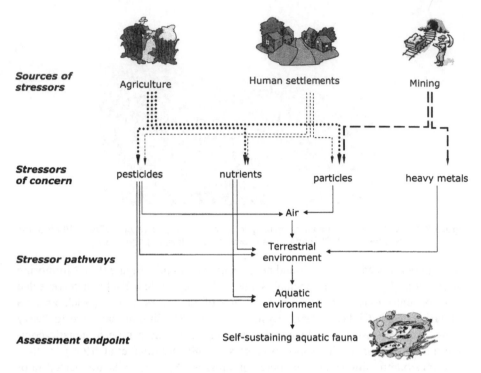

Figure 9.4   A simplified conceptual model relating sources of stressors to assessment end-points.

ANALYSIS PHASE

The analysis was based on the density of the stressor sources, the density of aquatic habitats, and the vulnerability of the endpoints to the different stressors. A detailed description of the methodological steps of the analysis phase can be found in Moraes et al. (2002).

Delimitation of Subareas

The geographical boundaries of the study are the entire catchment areas of the three main rivers crossing PETAR, which together drain an area of approximately 1000 km². These three rivers, Betari, Iporanga, and Pilões, are tributaries to the Ribeira River, which flows into the Atlantic Ocean. In addition to PETAR, the study area partially includes three other protected areas: Parque Estadual Intervales, Area de Proteção Ambiental Furnas, and Parque Estadual da Serra do Mar. The study area also includes four territories known as *quilombos*, areas owned by former slave communities. Maps on hydrology (IBDF 1987) and limits of *quilombos* and environmentally protected areas (ISA 1998) were compiled. The study area was subsequently divided into 14 subareas based on catchment limits, position in relation to the Ribeira River (headwaters or mouth), and inclusion in protected areas (Figure 9.6).

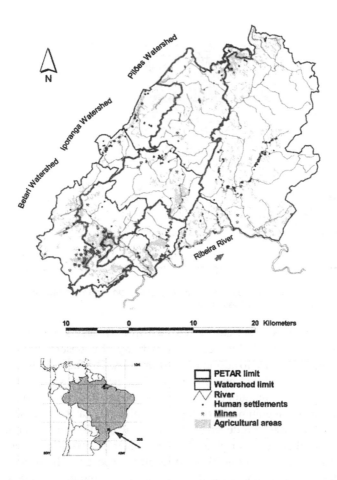

Figure 9.5   Map of watersheds of the three main rivers flowing though PETAR (Betari, Iporanga, and Pilões Rivers) showing the limits of the reserve and distribution of
human settlements, mines (limestone, lead, and gold), and subsistence or middle
intermediate scale agriculture.

Density of Sources

For the evaluation of the distribution of stressor sources in each area, maps on
mining activities (Cullen et al. 2001), human settlements (IBDF 1987), and vegetation coverage (SMA 1997) were compiled. A ranking system was designed based
on the number of sources per area (mines or households) or percentage of agricultural
land (e.g., percentage of bare soil plus anthropogenic field according to Secretaria
do Meio Ambiente do Estado de São Paulo, 1997). Ranks (0, 2, 4, 6) were given to
each source type, meaning no, low, medium, and high concentration, respectively.
The ranking criterion was based on an arbitrary division of the density of the source
scale into three equal parts, between zero and the maximum value.

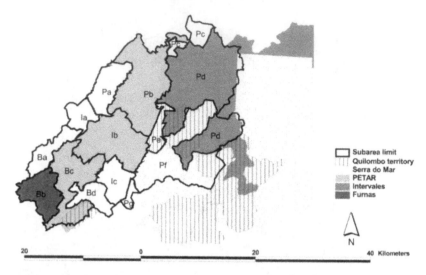

Figure 9.6   Environmentally protected reserves, *quilombo* territories and limits of the 14 subareas in the risk assessment.

The same type of stressor can be released from diverse kinds of sources but in different amounts, which influences the probability of exposure. Weighting factors for stressors were designed to account for differences. The first step was to determine which stressor was produced by each source. The next step was to assign weights, which would reflect the amount of stressor produced in comparison to other sources, representing no (0), medium (1) or high (2) release of a stressor. The weighting system for stressors was introduced by Moraes et al. (2002) as a further development of the exposure filter of the original RRM (Chapter 1). Exposure filters screen the source and habitat combinations likely to result in exposure. The stressor weighting system is designed to account for differences in the amounts of stressors emitted from the various sources, thus increasing the accuracy of the model.

Distribution of Aquatic Habitats

A ranking system (0, 1, 2) was designed based on the abundance of epigean and hypogean habitats in each subarea, meaning no, low, and high frequency, respectively.

Courses of streams were digitized based on hydrological maps (IBDF 1987). Due to the hilly topography and the high amount of precipitation, a large number of streams and rivers are present in the reserve and neighboring areas. For this assessment a relatively homogeneous distribution of number of rivers was assumed, and the rank 1 was assigned to all subareas.

More than 250 horizontal karstic caves and abyssal pits have been described in PETAR, and their occurrence is limited to limestone bedrock. A complete mapping of the distribution of the caves in the park and neighboring areas has not been performed. The ranking of the aquatic hypogean habitat was subsequently based on the percentage of limestone bedrock in each risk area according to the geological map of the area (Negri 1999). Ranks 0, 1, or 2 were given to each area, meaning no, low (1 to 25%), and high (26 to 50%) frequency, respectively, of limestone bedrock indicative of hypogean habitat.

Vulnerability of Endpoints to the Different Stressors

Epigean and hypogean rivers are interconnected in the area, but the characteristics of the habitat and the vulnerability of their fauna are quite distinctive. Hypogean streams typically have few tributaries, and the energy input is largely allochthonous (Trajano 1986). As a result of food scarcity, cave populations are usually small (Trajano 1986). The relatively low tolerance to chemical pollution (Mosslacher 2000), together with reduced population numbers, habitat restriction, and reduced geographical distribution, make troglobite (organisms restricted to a hypogean habitat) more vulnerable to ecosystem disturbances than epigean species (Trajano 1986). Once declined, recovery of a troglobite population is slow, and habitat re-colonization by dispersion is limited to contiguous subterranean ecosystems.

Weighting factors for effects were designed to evaluate the likelihood and extent to which exposure to different stressors may harm the habitat as a function of not only the system sensitivity, but also its ability to adapt to new conditions. Higher values represent higher probability of adverse effects. The criteria followed the assumptions that cave fauna is generally more susceptible to pollutants than surface fauna (Mosslacher 2000). In both habitats effects of pesticides were considered to be more severe than of metals due to the acute and unpredictable exposure of aquatic organisms after their application in agricultural fields.

Calculation of Relative Risk

Risk is an integration of two factors: the likelihood of the endpoint to be exposed to stressors and the likelihood of the stressor to cause undesired effects to the endpoint. The relative risk per area was calculated separately for the epigean and hypogean habitats based on Landis and Wiegers's model (Chapter 2 of this volume.) The modification was the introduction of a weighting factor for stressors, reflecting the relative amounts of contaminants released by the sources. All equations can be found in Moraes et al. (2002).

RISK CHARACTERIZATION

In the risk characterization, relative risks per source, per subarea, and per stressor were calculated using the outcomes of the analysis phase. High ecological risk here means a greater relative potential for adverse effects on the fauna living in rivers. The numbers generated by the risk estimations do not have a specific unit and should be used only for comparisons among identified stressors, subareas, or sources in this case.

Relative Risk per Source

The stressor sources of major concern were agriculture and human settlement (Figure 9.7) because both are abundant in the area. In addition, they may release both

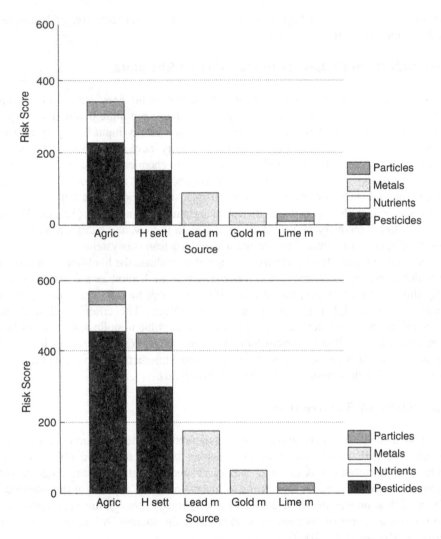

Figure 9.7   Relative risk per source to the epigean (above) and hypogean aquatic environment (below) on a relative scale. Agric (agriculture), H sett (Human settlement), Lead m and Gold m (inactive lead and gold mines with open and subterranean quarries) and Lime m (active and inactive limestone quarries). Stressors in legend at right.

pesticides and nutrients. Piles of waste rock from abandoned gold and lead mines might release heavy metals. However, they do not represent as great a threat as agriculture and human settlements, because the mines are concentrated to few subareas. Limestone mining is a comparatively minor threat as there are few limestone mines in the region and because the major pollutant of that activity is calcareous particles.

Relative Risk per Subarea

The relative risks per subarea to the aquatic fauna are presented in Figure 9.8 and Figure 9.9. The Betari catchment area (B) has a greater risk of being adversely

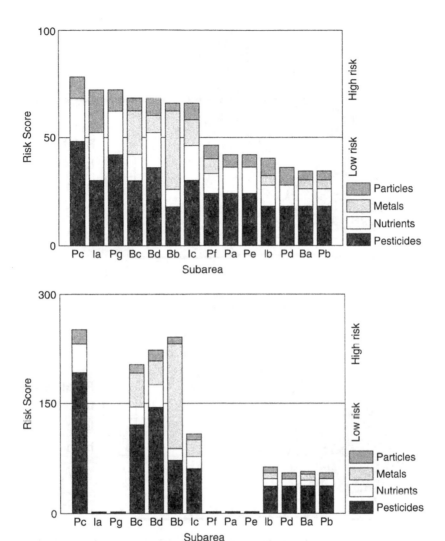

Figure 9.8   Relative risk per subarea to epigean (above) and hypogean aquatic fauna (below) on a relative scale. Stressors indicated in legend at right. Refer to Figure 9.6 for subarea designation. B = Betari, I = Iporanga and P = Pilões.

impacted as a result of human activities than the Pilões (P) and Iporanga (I) watersheds. Within the catchment area of Betari, both epigean and hypogean aquatic fauna of Bb, Bc, and Bd are at higher relative risk. This is due to the higher concentration of sources leading to a greater risk for exposure to both epigean and hypogean habitats.

For the subarea located within the environmentally protected area of Furnas (Bb), heavy metals stemming from the nonactive lead mine are of main concern. The subareas Bd and Bc are most likely to be impacted due to releases of pesticides and domestic sewage.

Another area of higher risk for both epigean and hypogean fauna is the northern headwaters of the Pilões River (Pc). Many houses are located in the area, and a large

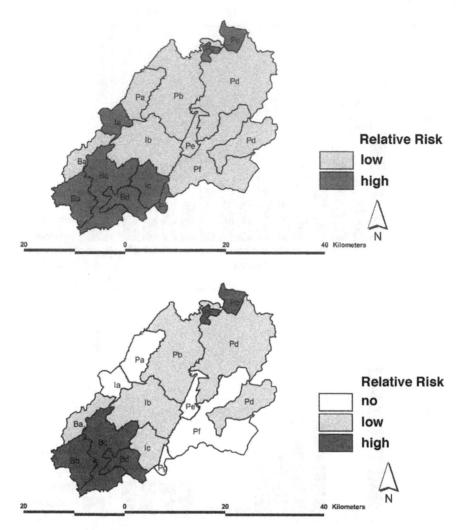

Figure 9.9 Relative risk per subarea to the epigean (above) and hypogean aquatic fauna (below).

percentage of the land of Pc is used for agriculture, mainly tomato fields that require application of pesticides. In addition, the use of fertilizers and erosion processes are likely to increase the nutrients and particles contents in the watercourses. Within the Pilões catchment area, Pg, the most downstream portion of the Pilões catchment area, is also of concern because of the exposure of epigean aquatic fauna to increased concentrations of nutrients and pesticides.

Inside the Iporanga catchment area, two areas (Ia and Ic) have a higher risk to epigean aquatic fauna, but low or no risk to hypogean organisms because the areas have little or no limestone bedrock. Subarea Ia includes the headwaters of the Iporanga River, where the highest concentration of households within the entire study area is found. In addition, two active limestone mines are present. Pesticides, nutrients, and particles are the stressors of concern. Small farms occupy nearly 15% of the Ic subarea, located in the lowest part of the river. The city of Iporanga, with a population of 2500 inhabitants, is situated in Ic near the confluence of the Iporanga and Ribeira Rivers. Subareas Ba, Ib, Pa, Pb, Pd, Pe, and Pf present relatively low or no risk to both epigean and hypogean aquatic fauna.

Uncertainty Analysis

There is a considerable degree of uncertainty in the risk estimations at a regional level. Part of the uncertainty is due to the simplifications made by the RRM, which disregards, for instance, degradation and fate processes, and the temporal variation in exposure.

Additional uncertainty related to simplifications in the model relates to the fact that it assumes (1) equal distribution of sources within the subareas, (2) equal distribution of habitats within the subareas, (3) correlation between the probability of exposure and the abundance of sources, (4) correlation between the probability of exposure and the abundance of habitats, (5) no transport of stressors between areas, and (6) absence of antagonistic and synergistic effects among stressors.

Another kind of uncertainty is related to the input of values in the model. For instance, a unique value was given as a weighting factor for effects of pesticides in epigean habitats, even though the term pesticide is used for a wide range of substances with distinct toxicological properties. Additionally, the study was forced to rely on out-of-date maps. Gathering site-specific, up-to-date data can reduce this kind of uncertainty.

DISCUSSION

The evaluation of ecological risks over large geographical areas requires RRAs. In these circumstances the applicability of traditional site-specific risk assessment procedures may have limited use. A few models have been suggested for that purpose (Graham et al. 1991; Clifford 1995; Gordon and Majumder 2000; Hartwell 1997) and all face constraints in terms of data availability, because regional-specific data are often not available and the cost of collecting new data at larger scales is normally high, especially in developing countries. The merits of the RRM applied in this RRA are not only related to low demand for input data, and consequently low cost, but also to transparent assessment models, plus the fact that its outputs are easy to use in risk communication and for creating priorities for management actions.

The RRA in PETAR yielded applicable products. Graphs and maps generated can be used for an effective communication of risk to stakeholders. The outcome of this assessment can assist the park management as regards risk mitigation in the area, for instance, dedicating available resources to reducing higher risks in the areas highlighted in this study. Results of the regional risk analysis could be incorporated in the new management plan of the park, which is currently being designed by members of the Forest Institute and PETAR administration and by an advisory council made up of representatives of local residents, governmental and nongovernmental agencies, tourism operators, researchers, and other experts.

Another important outcome of the RRA in PETAR was the generation of several testable hypotheses, for instance: (1) ecological effects are more likely to be observed in aquatic fauna living in subarea Bc compared to aquatic fauna living in subarea Ba; and (2) effects eventually observed in the aquatic fauna of subarea Bc are more likely to be caused by exposure to heavy metals and pesticides than by exposure to particles and nutrients. Some of these hypotheses guided site-specific investigations for local risk assessments which are reported in Moraes (2002) and Moraes et al. (2003). The studies at local scale included assessment of the baseline condition of the ecosystem and its natural range of variation, and detections of effects on different levels of biological organization. The risk characterization incorporated the weight-of-evidence ecoepidemiological approach for evaluation of causal relationships between exposure and effects. The outcome of the risk assessment at local scale showed important limitations of the RRM, which disregards transport of stressors between subareas. The results indicated that the model needs to be further developed and that exposure estimations in subareas downstream of the sources may be underestimated by the model.

All landuse problems mentioned in this study can also be found in many of the Brazilian reserves. There are 785 federal and state protected areas and Private Natural Heritage Reserves, which total 69,174,600 ha, or 8.13% of the country. According to the World Wildlife Foundation (1999), in 41% of the officially protected areas, more than 50% of the surroundings is already deforested. A large proportion (22%) of the reserves has problems regarding landuse in neighboring areas (50% of surrounding areas are occupied by agriculture, industrial, mining activities, or urban centers), and 12% of the reserves is influenced by systematic resource exploitation in more than 10% of their area. The merits of the relative risk assessment approach — low demand for input data, transparent assessment models, low cost, plus the fact that it is easy to use in risk communication — as shown by this case study — imply that the model can be used in other reserves in Brazil or even in other tropical areas in developing countries in the establishment of conservation priorities.

ACKNOWLEDGMENTS

The research for this chapter was financed by the Swedish International Development Agency (Sida).The authors would like to thank Gene Hoerauf for help with the geographical information systems. We are grateful to the PETAR administration for the logistic support of the project, and to Eleonora Trajano, Helio Shimada, and Martinus Filet who provided the background data.

REFERENCES

Bekerman, H.E. and Rabeni, C. 1987. Effects of siltation on stream fish communities, *Environ. Biol. Fish.*, 18, 285–294.

Beltman, D. et al. 1999. Benthic invertebrate metals exposure, accumulation, and community-level effects downstream from hard-rock mine site, *Environ. Toxicol. Chem.*, 18, 299–307.

Clifford, P.A. 1995. An approach to quantify spatial components of exposure for ecological risk assessment, *Environ. Toxicol. Chem.*, 14, 895–906.

Companhia de Tecnologia de Saneamento Ambiental (CETESB). 1988. Iguape-Cananéia-Vale do Ribeira: caracterização ecológica e avaliação toxicológica da população da Bacia, Technical Report, Secretaria do Meio Ambiente do Estado de São Paulo, São Paulo, Brazil.

Companhia de Tecnologia de Saneamento Ambiental. 1991. Avaliação da qualidade do Rio Ribeira do Iguape — Considerações Preliminares, Secretaria do Meio Ambiente do Estado de São Paulo, São Paulo, Brazil.

Cullen, L. et al. 2001. Agroforestry benefit zones: a tool for the conservation and management of Atlantic forest fragments, Sao Paulo, Brazil, *Natural Areas J.*, 21, 346–356.

Eysink, G.G.J. et al. 1988. Metais pesados no Vale do Ribeira e em Iguape Cananéia, *Ambiente*, 2, 6–13.

Galetti, M. and Fernandez, C. 2002. Palm heart harvesting in the Brazilian Atlantic rain forest: Changes in industry structure and illegal trade, *J. Appl. Ecol.*, 35, 294–301.

Gordon, S. and Majumder, S. 2000. Empirical stressor–response relationships for prospective risk analysis, *Environ. Toxicol. Chem.*, 19, 1106–1112.

Graham, R. et al. 1991. Ecological risk assessment at the regional scale, *Ecol. Appl.*, 1, 196–206.

Hartwell, S.L. 1997. Demonstration of a toxicological risk ranking method to correlate measures of ambient toxicity and fish community diversity, *Environ. Toxicol. Chem.*, 16, 361–371.

Instituto Brasileiro de Geografia e Estatística (IBDF). 1987. Mapa Topográfico das Regiões Sudeste e Sul do Brasil, Folhas Iporanga, Mina do Espírito Santo e Ribeirão Itacolomi. Instituto Brasileiro de Geografia e Estatística (IBGE), São Paulo, Brazil.

Instituto Florestal (IF). 1985. Plano conceitual de manejo para o Parque Estadual Turístico do Alto Ribeira – Projeto, Technical Report, Secretaria do Meio Ambiente, São Paulo.

Instituto Socioambiental (ISA). 1998. Projeto diagnóstico socioambiental do Vale do Ribeira. Documento Síntese — Anexo 1, Instituto Socioambiental, São Paulo, Brazil.

Karr, J.R. 1998. Rivers as sentinels: using the biology of rivers to guide landscape management, in *River Ecology and Management: Lessons from the Pacific Coastal Ecosystems*, Naiman, R.J. and Bilby, R.E., Eds., Springer-Verlag, New York, 502–528.

Landis, W.G. and Wiegers, J.A. 1997. Design considerations and a suggested approach for regional and comparative ecological risk assessment, *Hum. Ecol. Risk Assess.*, 3, 287–297.

Lino, C.F. 1992. Reserva da Biosfera da Mata Atlântica, Consórcio da Mata Atlântica, Universidade de Campinas, Campinas, Brazil.

Macedo, A.B. 2000. Dispersion halos associated with mineralization and mining pollution in the Ribeira River Valley, Paraná and São Paulo, Brazil, *Geosci. Dev.*, 6, 25–28.

Moraes, R. 2002. A Procedure for Ecological Tiered Assessment of Risks (PETAR). Doctoral Thesis, Environmental Systems Analysis, Chalmers University of Technology, Gothenburg, Sweden.

Moraes, R., Landis, W.G., and Molander, S. 2002. Regional risk assessment of a Brazilian rain forest, *Hum. Ecol. Risk Assess.*, 8, 1779–1803.

Moraes, R. et al. 2003. Establishing causality between exposure to metals and effects on fish, *Hum. Ecol. Risk Assess.*, 9, 149–169.

Mosslacher, F. 2000. Sensitivity of groundwater and surface water crustaceans to chemical pollutants and hypoxia: implications for pollution management, *Arch. Hydrobiol.*, 149, 51–66.

Myers, N. et al. 2000. Biodiversity hotspots for conservation priorities, *Nature*, 403, 853–858.

Negri, F.A. 1999. Mapa compilado de unidades litoestratigráficas de parte do Vale do Ribeira. Anexo Geo-01, in Contribuição ao Conhecimento do Meio Físico do Parque Estadual Turístico do Alto Ribeira - PETAR (Apiaí e Iporanga, SP), Price, F., Brix, K., and Lane, N., Eds., Instituto Geológico, Secretaria de Estado do Meio Ambiente, São Paulo, Brazil.

Petersen, R.J. 1986. Population and guild analysis for interpretation of heavy metal pollution in streams, in *Community Toxicity Testing*, STP 920,Cairns, J., Ed., American Society for Testing and Materials, Philadelphia, PA, 180–198.

Poulson, T.L. and Lavoie, K. 2000. The trophic basis of subsurface ecosystems, in *Subterranean Ecosystems, Ecosystems of the World*, Vol. 30, Wilkens, H., Culver, D.C., and Humphrey, W., Eds., Elsevier Science, Amsterdam, 231–250.

Sanchez, L. 1984. Caves and karst of the Upper Ribeira Valley, SP: a proposal of preservation, *Espeleo-tema*, 14, 9–21.

Secretaria do Meio Ambiente do Estado de São Paulo (SMA). 1991. Projeto PETAR — Parque Estadual Turístico do Alto Ribeira, Technical Report, Governo do Estado de São Paulo, São Paulo, Brazil.

Secretaria do Meio Ambiente do Estado de São Paulo (SMA). 1997. Macrozoneamento do Vale do Ribeira, Governo do Estado de São Paulo, São Paulo, Relatório Técnico, 019-SMA-MCZ-RT-P1400.

Swedish Environmental Protection Agency (SEPA). 1993. Eutrophication of Soil, Fresh Water and Sea, Report Number 4244, Stockholm, Sweden.

Shimada, H., Burgi, R., and Silva, M.H.B. 1999. Diagnóstico da mineração no PETAR e nas suas vizinhanças, in Contribuição ao Conhecimento do Meio Físico do Parque Estadual Turístico do Alto Ribeira— PETAR (Apiaí e Iporanga, SP), Shimada, H., Ed. Instituto Geológico, Secretaria de Estado do Meio Ambiente do Estado de São Paulo, São Paulo, Brazil, 170–223.

Trajano, E. 1986. Vulnerabilidade dos trogóbios a perturbações ambientais, *Espeleo-tema*, 15, 19–24.

Trajano, E. 2000. Cave fauna in the Atlantic tropical rain forest: composition, ecology, and conservation, *Biotropica*, 32, 882–893.

U.S. Environmental Protection Agency (USEPA). 1998. Guidelines for ecological risk assessment, EPA/630/R-95/002F, Risk Assessment Forum, U.S. Environmental Protection Agency, Washington, D.C.

United Nations Educational Scientific and Cultural Organization (UNESCO). 2001. Convention Concerning the Protection of the World Cultural and Natural Heritage Centre, Paris, France, available at http://whc.unesco.org/world.he.htm (accessed June 30, 2004).

World Wildlife Foundation (WWF). 1999. Áreas Protegidas ou Áreas Ameaçadas? Relatório da WWF sobre o Grau de Implementação e Vulnerabilidade das Unidades de Conservação Federais Brasileiras de Uso Indireto, Brasília, Série Técnica I.

CHAPTER 10

Using the Relative Risk Model for a Regional-Scale Ecological Risk Assessment of the Squalicum Creek Watershed

Joy C. Chen and Wayne G. Landis

CONTENTS

1-56670-655-6/04/$0.00+$1.50

PART I: USING THE RELATIVE RISK MODEL
FOR A REGIONAL-SCALE ECOLOGICAL RISK ASSESSMENT
OF THE SQUALICUM CREEK WATERSHED

Introduction

Ecological risk assessment (EcoRA) methodologies are well established, and general guidelines are listed in the "Guidelines for Ecological Risk Assessment" (USEPA 1998). Most EcoRA methods follow the three-phase approach: problem formulation, risk analysis, and risk characterization. These methods differ mostly in the risk analysis and the risk characterization phases. While many risk analysis and risk characterization methods are available (Landis et al. 1998), most of these methods are exposure- and effect-based methods that cannot accurately convey risks unless information is available for all exposure pathways for the risk components. Uncertainty associated with these methods increases greatly when there is insufficient exposure and effect data. As in most regional-scale assessments, there is insufficient information in this study to use the exposure- and effect-based methods. Subsequently, we used the alternative method, the ranked-based method for this study. The rank-based method is a probability-based method that determines the relative risks associated with each stressor instead of determining the absolute effects due to particular stressors. In cases where data are limited such as in this study, the rank-based method can minimize the uncertainties associated with the insufficient information on the characterization of exposure and ecological effects in the exposure–effect methods.

In this study, we followed the traditional three-phase approach of the EcoRA. We used the relative risk model (RRM), a ranked-based method, in the risk analysis phase of this EcoRA. We performed an EcoRA of the Squalicum Creek watershed, Bellingham, WA, using the RRM. The objective of our project is to determine the relative contribution of risks of adverse impacts of stressors to the Squalicum Creek watershed habitats, and to determine the utility of the RRM on a small-scale ecological system relative to the studies mentioned above.

Methods

Methodology used in this study was similar to that used by Landis and Wiegers (1997) and Wiegers et al. (1998) with few deviations from the original RRM in the risk analysis phase as stated below.

The risk analysis phase in the original methodology includes two steps: (1) performing a comparative analysis to determine the relative risks in each risk region, and (2) performing quantitative analyses to determine the severity of risk in the study area and to confirm the results from the comparative analysis. In this study, we only included the comparative analysis and left out the quantitative analysis. This is due to the limited site-specific quantitative data available for our study area, which is required by the quantitative analysis.

In addition to the risk components included in the original methodology, we have also included an extra risk component, the stressor group. We included these groups of stressors in this study to indicate the possible types of stressors releasing or resulting from the stressor sources.

Possible endpoint location is another extra risk component apart from those listed in the original methodology. We included the possible endpoint location because we included abiotic endpoint in our study. The geological information of the endpoints is essential for a risk assessment. The location of biotic endpoints is normally defined by the habitat of the biotic endpoints. However, the location of abiotic endpoints does not necessarily correlate with any type of habitat and, therefore, using the habitat to define these endpoint locations is improper. Therefore, we added a new risk component, the possible endpoint location, to better represent the abiotic endpoint location. Extra filters have also been added to this study in response to the additional risk components.

In the original methodology, risk scores for each risk region were calculated by multiplying the risk ranks by the list of associated filters, called the weighting factor. Risks resulting from a particular source and occurring in a particular habitat were calculated by adding the related score for each risk region. In this study, we modified the basic equations to account for the abiotic endpoints and the alterations in the filters in this study.

PROBLEM FORMULATION

This section summarizes the physical and biological characteristics of the study area, identifies the stressors and endpoints derived from stakeholders' values, defines risk regions, and includes the site conceptual model.

Study Area

The Squalicum Creek watershed lies within the city of Bellingham and extends into the unincorporated areas of Whatcom County (Figure 10.1). The study area includes the entire Squalicum Creek watershed plus the portion of the Port of Bellingham landfills into which the creek drains. The landfills were included for two reasons: (1) the landfills could potentially act as a physical barrier to migratory fish in and out of the creek, and (2) the stormwater from these landfills flows directly into the mouth of the creek.

The study area is 62 km² and the creek measured 5.99 km from the longest tributary to the outfall where it drains into the bay. The hydrology system is comprised of the main stream, Squalicum Creek, and a main tributary, Baker Creek (Figure 10.1). The entire system generally flows from northeast to southwest. There are two constructed lakes, Sunset Pond and Bug Lake, located in the middle section of Squalicum Creek.

For this assessment, the study area was divided into six risk regions (Figure 10.2). Region boundaries were defined by grouping parcels with similar landuse types, topography (USGA 2000), and hydrology (Hoerauf 1999). In cases where these factors were insufficient to determine the boundaries, the city boundary was followed.

Regions 1 and 3 are located within the city limits, regions 4, 5, and 6 are located in the county, and region 2 is under the jurisdiction of both the City of Bellingham

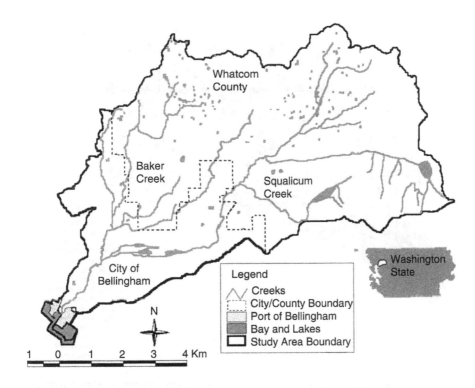

Figure 10.1 Study area boundary for the Squalicum Creek watershed ecological risk assessment.

and Whatcom County. Region 1 consists of the Port of Bellingham, along with mainly residential, mining, transportation, and park landuse. It contains the lower portion of Squalicum Creek that receives water from all tributaries. Region 2 is comprised mainly of commercial, mining, heavy industrial, agricultural, and undeveloped landuse. It contains one natural lake, two constructed lakes, and the middle section of both Baker and Squalicum Creeks. Region 3 is comprised mainly of commercial and residential landuse, along with a golf course and some undeveloped land. It contains the middle portion of Baker Creek. Region 4 consists mainly of forested, undeveloped, agricultural, and residential landuse. It contains two natural lakes and a portion of the Squalicum Creek headwaters. Region 5 consists of mainly agricultural, residential, and forested landuse. It also contains a portion of the Squalicum Creek headwaters. Region 6 consists of mainly agricultural, residential, forested, and undeveloped landuse. It contains the upstream sections of Baker Creek.

Ecological Endpoints Identification

The ecological endpoints were chosen by members of the Squalicum Creek Risk Assessment Group that consists of stakeholders such as the City of Bellingham, Whatcom County Conservation District, and the Nooksack Salmon Enhancement

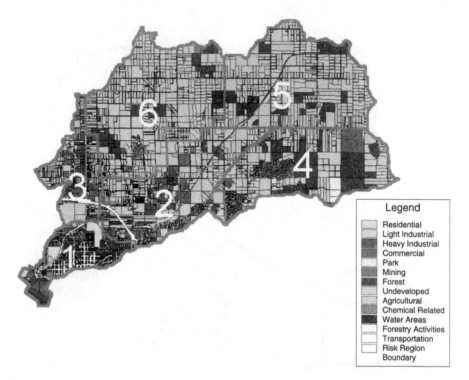

Figure 10.2   Risk regions and landuses in the study area. (See color insert following page 178.)

Association. The USEPA Guidelines for Ecological Risk Assessment (USEPA 1998) were followed in selecting the assessment endpoints. The criteria for endpoints are: (1) ecological relevance, (2) susceptibility to known or potential stressors, and (3) relevance to management goals. The first two endpoints are classified as abiotic endpoints and the last four are classified as biotic endpoints. The assessment endpoints for this assessment are:

1. Abiotic endpoints
   • Flood control
   • Adequate land and ecological attributes for recreational uses
2. Biotic endpoints
   • Viable nonmigratory coldwater fish populations
   • Life cycle opportunities for salmonids
   • Viable native terrestrial wildlife species populations
   • Adequate wetland habitat to support wetland species populations

Conceptual Model

The assumed relationships among the stressor sources, stressors, habitats, and endpoints for the study area are summarized in the conceptual model (Figure 10.3). This model serves as the basis for all risk assessment calculations discussed in the following sections.

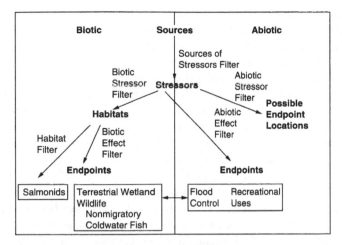

Figure 10.3   Conceptual model for the Squalicum Creek watershed ecological risk assessment.

RISK ANALYSIS

In general, we followed the risk analysis methodology used by Wiegers et al. (1998) with minor deviations as previously discussed.

Identifying and Ranking

We identified and ranked each stressor source, possible endpoint location, and habitat. We divided each of these risk components into four groups: no, low, medium, and high concentration and we assigned ranks 0, 2, 4, and 6 to each group, respectively. The no concentration group equals 0% of the risk component in a risk region. For example, if there were no warmwater habitat available in risk region 1, then a risk rank of 0 would be assigned to the warmwater habitat in risk region 1. The group intervals were categorized using Jenk's Optimization in ArcView® GIS. This ranking method was applied to all risk components except for coldwater fish habitat.

Stressor Sources

Eleven landuses were classified as the sources of stressors. They are: agricultural, residential, light industrial, heavy industrial, mining, chemical industries, commercial, park, transportation, forestry activities, and stream barrier construction. Stream barrier construction landuse is defined as the construction of any physical object such as a culvert that could inhibit the migration of aquatic species. Landuse categories were determined using the following sources: (1) the Whatcom County Code (Whatcom County Council 2000) and the Whatcom County Land Use Codes (Whatcom County Assessors Office 2000) provided by the Whatcom County Assessors' Office, (2) assistance from the City of Bellingham Planning Office, (3) USEPA WRIA BASINS database (USEPA 2000), and (4) fish presence mapping project data (Whatcom Conservation District 2000).

Eight stressor groups were chosen for this study because they could potentially adversely affect the endpoints. These stressor groups are: increased runoff, increased chemicals, altered stream flow, increased nutrients, altered forest, altered wetland, increased sediments, and introduced terrestrial foreign species.

Increased runoff was considered a stressor because it can increase the peak flow and soil erosion. It can also decrease the subsurface flow and, therefore, decrease the amount of water available for the species. Increased chemicals were identified as a stressor because they can lead to toxicity. An alteration of stream flow could change the stream temperature, obstruct the migratory routes for aquatic species, alter the water quality, and change the composition of the substrate, i.e., the aquatic habitat. Increasing the amounts of nutrients such as fecal coliform, nitrogen, and phosphorous compounds can lead to oxygen depletion in the aquatic habitat. Alterations of the forests and wetlands were considered as stressors because they reduce habitat availability to species. Alteration of wetlands could decrease the vegetation cover along the streams and lakes and, therefore, increase the water temperature and decrease the pool habitats and nutrients in the system. Altering the wetlands could also change the soil and water chemistry in the watershed and in the adjacent marine habitat. Increased sediment was identified as a stressor because it could reduce the amount of sunlight penetrating through the water, thereby reducing the photosynthesis process. Increased sediment could also disrupt the oxygen intake of some aquatic species and threaten their survival. Bringing in terrestrial-introduced species could lead to potential competition with the native species for resources and habitats. A summary of the assumed relationships between the sources of stressors and the stressor groups is indicated in Figure 10.4.

All landuses but mining and stream barrier construction were ranked using the percentage of land coverage of each landuse per region. The number of mines and stream barriers was used to rank the mining and the stream barrier construction landuse, respectively. Transportation landuse coverage was determined using two sources: the landuse parcel GIS data that include the concentration of all transportation facilities except roads, and the City of Bellingham GIS street data that include the area of the street coverage. Forestry activities were found only in region 4, and a low rank was assigned due to the relatively small land coverage of these activities. Table 10.1 provides a summary of the criteria for the stressor source ranks.

Habitats

For this assessment, all areas with saline water were included as coastal habitat. Lakes with surface area greater than 139.5 m^2 defined the warmwater habitat. Coldwater fish habitat included all streams plus lakes with surface area less than 139.5 m^2. Riparian habitat included areas within 60.96 m from the streams and lakes that were classified as the following landuses: forested, undeveloped, and park. Terrestrial habitat included all areas other than the riparian habitat that were classified as forested, undeveloped, or park landuse.

All but the coldwater fish habitat ranks were determined using the methodology described in the identifying and ranking section. The coldwater fish habitats were assumed to be of good quality and were assigned a high rank for all regions due to

Legend

| x | Pathway Exists |
| (shaded) | Pathway Absent |

		Stressors							
		Increased Runoff	Increased Sediments	Increased Chemicals	Introduced Terrestrial Foreign Species	Altered Stream Flow	Increased Nutrients (N/P/Fecal Coliform)	Altered Forest	Altered Wetland
Landuse (Sources of Stressors)	Agricultural	x	x	x	x	x	x	x	x
	Residential	x		x	x	x	x	x	x
	Light Industrial	x		x		x	x	x	x
	Heavy Industrial	x		x		x	x	x	x
	Mining	x	x	x		x	x	x	x
	Chemical Industries	x		x		x		x	x
	Commercial	x		x		x	x	x	x
	Park	x	x	x		x	x	x	x
	Transportation	x		x		x		x	x
	Forestry Activities	x	x	x		x	x	x	x

Figure 10.4   Assumed relationships between stressor sources and stressor groups.

the following reasons: (1) there are insufficient water quality and habitat data for the creek in all risk regions, (2) all regions include sections of the creek, and (3) there are insufficient data to determine the land coverage of the creek. Coastal habitat was found only in region 1 and was assigned a high rank. Table 10.2 provides a summary of the habitat ranks criteria.

Possible Endpoint Locations

Areas with park landuse defined possible recreational uses endpoint location for this risk assessment. The 200-year floodplain for the Squalicum Creek watershed defines the possible flood control endpoint location. The percentage of the possible endpoint locations in each risk region was used to determine the ranks. Table 10.3 provides a summary of the criteria for possible endpoint location ranks.

FILTERS

Six filters were used in this assessment to represent the relationships among the risk components. The sources-of-stressors filter indicates if a particular source releases a certain stressor group. The biotic stressor filter indicates if a stressor would occur and persist in and affect the habitat. The biotic effect filter indicates if an alteration of the habitat could affect an endpoint. The habitat filter for salmonids indicates if the streams in a particular risk region are located upstream of a physical barrier to salmonid migration. The habitat filter is included because of the unique

Table 10.1   Ranking Criteria for Stressor Sources

Landuses	Criteria	Ranks
Agricultural	% Agricultural	
	0	0 (No impact)
	0.76–12.83	2 (Low)
	12.84–22.37	4 (Medium)
	22.38–34.93	6 (High)
Residential	% Residential	
	0	0 (No impact)
	21.82–24.72	2 (Low)
	24.73–29.71	4 (Medium)
	29.72–42.84	6 (High)
Light industrial	% Light industrial	
	0	0 (No impact)
	0.01–0.29	2 (Low)
	0.30–0.67	4 (Medium)
Heavy industrial	% Heavy industrial	
	0	0 (No impact)
	0.01–0.37	2 (Low)
	0.38–0.97	4 (Medium)
	0.98–5.45	6 (High)
Mining	Number of mines	
	0	0 (No impact)
	1–2	2 (Low)
Chemical industrial	% Chemical industrial	
	0	0 (No impact)
	0.001–0.01	2 (Low)
	0.011–0.50	4 (Medium)
Commercial	% Commercial	
	0	0 (No impact)
	0.31–0.41	2 (Low)
	0.42–11.36	4 (Medium)
	11.37–29.73	6 (High)
Park	% Park	
	0	0 (No impact)
	0.1–0.45	2 (Low)
	0.46–0.92	4 (Medium)
	0.93–10.4	6 (High)
Transportation	% Transportation	
	0	0 (No impact)
	0.73–1.1	2 (Low)
	1.2–5.44	4 (Medium)
	5.45–7.73	6 (High)
Forestry activities	% Forestry activities	
	0	0 (No impact)
	0.1–2.61	2 (Low)
Physical barrier construction	Number of physical barriers	
	0	0 (No impact)
	1	6 (High)

Table 10.2   Ranking Criteria for Stressor Groups

Habitats	Criteria	Ranks
Warm water	% Warm water	
	0	0 (No impact)
	0.01–0.03	2 (Low)
Cold water	Stream absent	0 (No impact)
	Stream present	6 (High)
Riparian	% Riparian	
	0	0 (No impact)
	2.72–3.61	2 (Low)
	3.62–5.11	4 (Medium)
	5.12–7.2	6 (High)
Terrestrial	% Terrestrial	
	0	0 (No impact)
	10.85–14.74	2 (Low)
	14.75–26.72	4 (Medium)
	26.73–38.07	6 (High)
Coastal	% Coastal	
	0	0 (No impact)
	0.1–10.33	2 (Low)

Table 10.3   Ranking Criteria for Possible Endpoint Locations

Possible Endpoint Locations	Criteria	Ranks
Recreational uses	% Park landuse	
	0	0 (No impact)
	0.1–0.45	10 (Low)
	0.46–0.92	20 (Medium)
	0.93–10.4	30 (High)
Flood control	% 200-year floodplain	
	0	0 (No impact)
	0.1–2.5	10 (Low)
	2.6–5.98	20 (Medium)
	5.99–8.86	30 (High)

migratory behavior of salmonids. The habitat filter enables us to address the specific portion of the habitat the salmonids utilize and assesses the impact of each physical barrier to salmonids. The abiotic stressor filter indicates if a stressor would occur and persist in the possible endpoint location. The abiotic effect filter indicates if the stressor could affect an endpoint. For all but the habitat filter for salmonids, if the answer to the questions is yes, which indicates the pathway exists, a rank of 1 is assigned. In cases where the answer is no, a rank of 0 is assigned. For the habitat filter for salmonids, a 1 is assigned if no stream in the region is located upstream of a physical barrier, a 0.5 is assigned if only portions of the streams in the region are located upstream of a barrier, and a 0 is assigned if all the streams in the region are located upstream of a barrier.

INTEGRATING RANKS AND FILTERS

By following the original methodology described by Landis and Wiegers (1997), we integrated the risk ranks and filters to generate risk scores. All equations in this study were derived from the basic equations used in their study as shown in Equations 10.1, 10.2, and 10.3 (Appendix A). Methodology used to calculate the risk scores in this study is listed in the following sections.

Endpoint Risk Scores

Endpoint risk scores signify the relative risks to each endpoint. Each endpoint risk score is a summation of all the risk scores contributing to the particular endpoint in the entire study area (Equation 10.4 through Equation 10.6 in Appendix A).

Stressor Risk Scores

Stressor risk scores indicate the relative risks contributed by each of the stressors. Each stressor risk score is a summation of all the risk scores contributed by the particular stressor in the entire study area (Equation 10.7 in Appendix A).

Stressor Sources Risk Scores

The stressor sources risk scores represent the relative risks contributed by each of the stressor sources. The risk score of each source is a summation of all the risk scores contributed by the particular stressor source in the entire study area (Equation 10.8 in Appendix A).

Habitat Risk Scores

Habitat risk scores indicate the relative risks occurring within a particular habitat. Each habitat risk score is a summation of all the risk scores contributed by the particular habitat in the entire study area (Equation 10.9 in Appendix A).

Risk Region Risk Scores

Risk region risk scores represent the relative risks to each risk region. Each risk region risk score is a summation of all the risk scores contributing to the particular risk region (Equation 10.10 in Appendix A). Jenk's Optimization was also performed to cluster the risk regions into high, medium, and low risk categories.

Risk Characterization

This section summarizes the information in the problem formulation phase and in the analysis phase to produce a list of risk estimation for the study area. This section describes the significance of the risk estimation in terms of stakeholders' values, determines the uncertainties, and lists the assumptions for this risk assessment. Assumptions for this risk assessment are the same as those listed in Landis and Wiegers (1997) and Wiegers et al. (1998).

RISK ESTIMATION RESULTS

We summarized the risk results from the risk analysis phase to generate a list of risk estimations. In the following sections, we state the risk estimation results associated with each risk component. At the end of these sections, we address the relevance of the risk estimations to the entire watershed. The risk estimation results only represent the relative probability of risks to each risk component and not the actual magnitude of risks. Using these risk estimations directly to quantify the magnitude of risks would be inaccurate due to the uncertainties associated with the risk assessment. It is necessary to integrate the risk estimations with site-specific quantitative data to accurately determine the magnitude of risks.

Stressor Sources

Table 10.4 shows a summary of the stressor sources results. The risk assessment indicated that residential landuse contributed the most risks to the watershed, whereas light and chemical industries, mining activities, forestry activities, and the construction of stream barriers contributed relatively less risks to the watershed. Results also showed that residential, mining, commercial, park, and transportation landuses contributed the most risks to region 1, while agricultural, light, heavy, and chemical industrial landuses contributed the most risks to region 2. Stream barrier construction and commercial landuse contributed the most risk to region 3. Forestry activities were observed only in region 4. The RRM results show that commercial landuse contributed more risks to region 1 than to region 3; however, due to the uncertainties associated with the model, the small risk differences between the two regions were considered insignificant.

Stressors

Table 10.5 shows a summary of the stressor results. The RRM indicated that stream flow alteration, altered forest, and altered wetland contributed the most risk to the watershed. Increased nutrients and introduced terrestrial foreign species contributed relatively less risks to the watershed. Results also showed that all stressors except increased runoff and introduced terrestrial foreign species contributed the most risks to region 1. Increased runoff and introduced terrestrial foreign species contributed the most risks to region 2 and region 4, respectively.

Habitats

Table 10.6 shows a summary of the habitat results. The assessment indicated that the coldwater habitat is at most risk and the warmwater habitat and the coastal habitat are at relatively small risk. Results also showed that warmwater habitat is

Table 10.4 Stressor Sources Ranks Result (numbers represent risk scores)

Risk Regions	Agricultural	Residential	Light Industrial	Heavy Industrial	Mining	Chemical Industrial	Commercial	Park	Transportation	Forestry Activities	Physical Barrier Construction
						Sources					
1	1168	3216	0	2064	1000	0	1784	3420	2676	0	0
2	1912	892	1688	2532	812	748	1496	908	1496	0	0
3	772	702	672	672	0	0	1764	1484	1176	0	744
4	762	1412	0	640	0	542	1084	696	542	696	588
5	1188	1080	0	0	0	0	284	0	284	0	0
6	632	876	0	264	0	456	228	0	228	0	0

Table 10.5 Stressor Group Ranks Result (numbers represent risk scores)

Risk Regions	Increased Runoff	Increased Chemical	Stream Flow Alteration	Increased Nutrients	Altered Forest	Altered Wetland	Increased Sediments	Introduced Terrestrial Foreign Species
				Stressors				
1	2220	2208	3000	1260	2940	3000	540	160
2	2280	1680	2340	864	2580	2340	256	144
3	1100	1100	1736	504	1540	1736	210	60
4	1020	1220	1274	588	1220	1274	168	198
5	480	480	416	312	480	416	108	144
6	520	520	360	216	520	360	48	140

Table 10.6   Habitat Ranks Result (numbers represent risk scores)

Risk Regions	Habitats				
	Warmwater	Coldwater	Riparian	Terrestrial	Coastal
1	0	3192	2816	392	888
2	704	3168	2784	768	0
3	0	2100	966	560	0
4	390	1950	2730	792	0
5	0	1764	832	240	0
6	0	1392	732	560	0

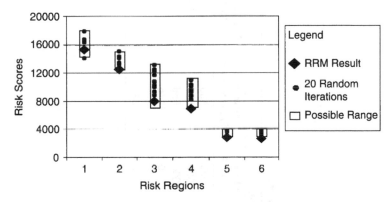

Figure 10.5   Sensitivity analysis: random component analysis result.

most at risk in region 2, and coastal habitat is only found in region 1. Coldwater habitat is most at risk in regions 1 and 2; riparian habitat is most at risk in regions 1, 2, and 4; and terrestrial habitat is most at risk in regions 2 and 4.

Endpoints

Table 10.7 shows a summary of the endpoint risk results. The RRM indicated that wetlands are the most at-risk endpoint in the watershed, and terrestrial wildlife species are the least at-risk endpoint. Life cycle opportunity for salmonids is the endpoint with the second lowest risk. Results also showed that nonmigratory cold-water fish, salmonids, and flood control endpoints are most at risk in regions 1 and 2, while terrestrial wildlife species are most at risk in regions 1 and 4, and recreational uses and wetlands are most at risk in region 1 and region 2, respectively.

Risk Regions

Figure 10.5 shows the risk results of risk regions as indicated by the RRM result. The risk assessment indicated that there is a strong risk gradient in which the risk decreases with the increasing risk regions number, i.e., the risk decreases as it moves from the downstream regions located in the city limits to the upstream regions that are located in the county area. Jenk's Optimization categorized regions 1 and 2 as high risks, regions 3 and 4 as medium risks, and regions 5 and 6 as low risks.

Table 10.7 Endpoint Ranks Result (numbers represent risk scores)

Risk Regions	Endpoints					
	Nonmigratory Cold-Water Fish	Life Cycle Opportunities for salmonids	Flood Control	Terrestrial Wildlife Species	Recreational Uses	Wetlands
1	1768	2064	3600	1196	4440	2260
2	1752	2104	3600	1080	1460	2488
3	1116	558	2000	556	2360	1396
4	1560	910	0	1176	1100	2216
5	796	796	0	328	0	916
6	940	0	0	524	0	1220

Relative Risk in the Squalicum Creek Watershed

This risk assessment determined that residential landuse is the source that contributes the most risks to the watershed. This result is not surprising knowing that 35% of the watershed consists of this landuse. The assessment also identified the stream flow alteration, altered forest, and altered wetland stressors as contributing the most risks to the watershed. This is due to the fact that these stressors could affect the endpoints through more exposure pathways than other stressors. Coldwater habitat was found to be most at risk in the Squalicum Creek watershed, especially in regions 1 and 2. Riparian habitat is the second most at-risk habitat, and it is most affected in regions 1, 2, and 4. Together these lead to the conclusion that the remediation of coldwater and riparian habitats in regions 1 and 2 should reduce the most risks to habitats in the watershed.

Wetland is the endpoint that is determined to be the most at risk. Wetlands are affected by the alteration of all habitats and, therefore, more exposure pathways are linked to this endpoint. Life cycle opportunity for salmonids receives relatively low risk because in three of the six risk regions, the construction of physical barriers prevented migration of salmonids to the upstream regions of these barriers. The absence of salmonids in these upstream regions leads to incomplete exposure pathways; therefore, salmonids are not at risk in these regions. The predicted strong risk gradient, increasing risk from upstream regions to downstream regions, is expected because there is a greater combination of habitats and stressor sources in the downstream regions than in the upstream regions.

UNCERTAINTY ANALYSIS

As mentioned earlier, we included an uncertainty analysis in the risk characterization phase to address all the uncertainties associated with this risk assessment. One of the sources of uncertainties in this risk assessment is stochasticity, which refers to the random nature of the universe, such as the random variations of endpoint responses to stressors. This type of uncertainty can only be estimated and usually cannot be reduced. Most uncertainties for this study are due to the lack of data or knowledge regarding the risk components and ecological pathways. Exposure pathways in this study were assigned based on professional judgment, which could potentially lead to error in the model. Predetermined risk components such as stressors and habitats could also lead to error in the risk predictions.

In this study, we assumed that risk in each region is discrete, but this assumption could lead to uncertainties because most stressors flow from upstream to downstream, and some risks from upstream regions could potentially enter into the downstream regions. Due to the lack of data, coldwater habitat in all risk regions was assigned a high rank for this study. There is a potential for variation in the coldwater habitat that could lead to uncertainty in the RRM result. Uncertainties also arise from the lack of information regarding the undeveloped land conditions. The undeveloped land makes up a portion of the terrestrial habitat; therefore, variation in the undeveloped land data could lead to variation in the terrestrial habitat data input into the

RRM. There are currently insufficient data for stressors, habitats, and possible endpoint locations in the Squalicum Creek watershed, especially in regions 3, 5, and 6. Landuse data were used in this study as a substitute for the missing information. The landuse data were not collected for use in the risk assessment; therefore, some landuse categories contribute uncertainties to the model due to inadequate description of the landuse types. For example, the undeveloped landuse varies greatly from forested land to open grasslands; this leads to potential variation in the habitat data input into the model. As mentioned earlier, there are uncertainties regarding the exposure pathways. For example, due to the increased use of retention ponds in recent years and insufficient data regarding pond location and efficiency, the completeness of exposure pathways between increased runoff and the commercial and industrial landuses is not clear. There are also uncertainties regarding the effects of seasonal patterns due to insufficient temporal data. The process of calculating risk estimates in the risk analysis phase also introduces uncertainties to the risk predictions due to model variance and possible model bias.

Sensitivity Analysis Methodology

Most of the uncertainties mentioned above are quantifiable, and we quantified them by performing a sensitivity analysis. In this study, we included six sensitivity analyses. They are categorized into four types of analysis: geographical, single component, exposure pathway, and random component. We first describe the methodology we used for each of the sensitivity analyses and then list the results and the significance of these sensitivity analyses. The purpose of the geographical analysis is to test the sensitivity of the model to upstream–downstream effects. We assumed a range of different percentages of risk from upstream regions to enter into the downstream regions to determine how this added risk would change the relative risk in the entire study area. We added 5 to 100% of the upstream regions' risks to the downstream regions at a 5% interval. For example, assuming that 10% of the risks from upstream regions would add to the risks in the downstream regions, then the total risk score for region 1 would equal region 1's risk score plus 10% of region 3's risk score plus 10% of region 2's risk score, where the total risk score for region 3 would equal region 3's risk score plus 10% of region 2's risk score plus 10% of region 6's risk score.

There are two separate analyses in the single-component analysis. In each of these analyses, a single risk component was altered in the RRM, and the risk results were compared to the original RRM result. The two risk components were coldwater habitat and terrestrial habitat. The single-component analysis removes the coldwater habitat from the RRM and the undeveloped landuse from the terrestrial habitat data to assess the sensitivity of the model to these habitats.

The exposure pathway analysis consists of two separate analyses. In both analyses, one or more exposure pathways were altered, and the risk results were compared to the original RRM result. The exposure pathway analysis removes the exposure pathways between increased runoff and the commercial and industrial landuses to assess the sensitivity of the RRM to these pathway uncertainties. The exposure pathway analysis also tested the model sensitivity to other exposure pathways by assuming all pathways to be complete.

The random component analysis randomized the possible ranges of stressor sources, habitats, and abiotic endpoint locations in 20 simulations to evaluate model bias. The possible ranges are defined by +/– 10% of the value of the risk components from the risk criterion breakpoint. For example, if the percentage of terrestrial habitat in region 5 is 14.74%, which is within 10% of the low and the medium risk criterion breakpoint (Table 10.2), then the possible range of risk score for the terrestrial habitat in region 5 would equal 2 and 4. The random component analysis results allow us to determine the model sensitivity. The sensitivity of the model involves two things: (1) ability of the model to differentiate the relative risks – high, medium, and low risk for the risk regions, and (2) ability of the model to generate nonbiased data, while the output data would correlate with the input data. For example, could the model produce random output data with random data input and discriminate high-risk regions when risk-related data are used?

The possible conditions of the risk regions are determined using similar methodology as in the random component analysis. Instead of randomizing the risk components within their possible ranges, the highest and the lowest possible risk combination of the risk components is used to represent the possible conditions of the regions.

Sensitivity Analysis Results

The geographical analysis indicated that the risk model is not sensitive to upstream–downstream effect below 20%. If 20% or more risk of the upstream regions was added to that of the downstream regions, risk region 2 would change from a high risk to a medium risk, while risks in all other regions remained the same. The single-component analysis shows that the alteration of a single component, the coldwater habitat and the terrestrial habitat, did not change the risk results of the regions. This indicates that variation of a single component is not likely to change the results of the risk regions; instead, the model is more sensitive to the combined effects from variations of multiple components. Both exposure pathway analyses indicated that pathway alteration does not change the risk results of the regions, showing that the model is less sensitive to the filters than to the other risk components. The random component analysis results did not show a risk pattern; randomized values in the model produced randomized results within the possible range (Figure 10.5). This indicates that the model does not produce biased results. Conditions of the possible risk regions are also shown in Figure 10.5. Taking conditions of the possible risk regions into account, there is still a clear risk trend. This is due to the fact that the possible risk range is limited by the data input into the RRM. Results indicated that region 3 has the greatest potential for variations, and regions 5 and 6 have the least potential. It also indicated that the RRM results for all regions except region 1 might be underestimated. Using Jenk's Optimization to rank all the possible combinations of the highest and lowest possible risk ranks reveals that region 2 can potentially change from a high risk to a medium risk. This result is consistent with the geographical analysis result. The Jenk's Optimization results also indicated that region 3 can potentially change from a medium risk to a high risk. Taking all the possible variations into consideration, risks in regions 1, 2, and 3 are consistently higher than regions 4, 5, and 6. In conclusion, the sensitivity results demonstrate the ability of RRM to differentiate the relative risks for the risk regions without producing biased results. The sensitivity results also show that the RRM is most sensitive to the input data.

DISCUSSION

In recent years, the level of concern for the Squalicum Creek watershed has elevated due to the increased adverse impacts of anthropogenic activities on the watershed. Various organizations have conducted studies in an attempt to assess the effects of these activities to the Squalicum Creek watershed. Although these studies can detect some of the adverse effects, most of these studies are only descriptions of the existing conditions, and they do not explain the relationship between these conditions and their sources. This assessment fills this information gap by integrating the effects of multiple individual decisions on a regional-scale ecosystem. It demonstrates our ability to determine the relative contributions of risks of adverse impacts of multiple (biological, chemical, and physical) stressors on the Squalicum Creek watershed and shows the suitability of the application of the RRM to the watershed management process. This assessment also demonstrates the applicability of the RRM to a relatively small-scale ecological system and illustrates its potential in assessing risks in similar ecosystems.

Application of the Relative Risk Model

The application of the RRM to the Squalicum Creek watershed was successful; only a few modifications to the model were necessary. As in the original RRM, stakeholders were involved in the Squalicum Creek watershed risk assessment process, and assessment endpoints were defined according to their values. The original RRM methodology for defining risk regions was successfully applied to this study. A conceptual model was developed and used as a basis for the risk calculations, and the basic RRM ranking concept was followed in this assessment. Sources and habitats were identified and ranked, and filters were incorporated into the risk calculations to determine the final risks for endpoints and risk regions.

The terminology in this study differs slightly from the original RRM. An additional risk component, the groups of stressors, was added to this risk assessment; this new risk component is equivalent to the sources in the original RRM. This change was made because there were insufficient stressor data for the Squalicum Creek watershed. As a result, landuse data were used to replace the stressors data. Another change made to the model was the addition of filters. To accommodate the addition of the abiotic endpoints and the extra risk component, the stressor groups, the original exposure and effect filters were split into five separate filters: the sources of stressors filter, the biotic stressor filter, the abiotic stressor filter, the biotic effect filter, and the abiotic effect filter. The habitat filter was also added to account for the effects of the physical stressors on the exposure to migratory species. The addition of the risk component, possible endpoint locations, was also necessary to address the geographical information of the abiotic endpoints. The abiotic filter rank numbers also differ from the original RRM so the abiotic endpoint risk ranks are comparable to the biotic endpoint risk ranks. The ranking method differs slightly from the original RRM; instead of breaking the stressors and habitats categories into equal divisions, this model divided them using natural breaks. In summary, most changes to the original RRM were due to the addition of the abiotic endpoints and the lack of stressors information for the Squalicum Creek watershed.

Risk Management

In all cases, the results from this assessment show that the risk to the watershed can be minimized by lowering the number of stressors and isolating the habitats from these stressors instead of just increasing habitats in these regions. Increasing habitats in these regions without reducing the amount of stressors would only lead to greater risks in these habitats, because exposure would be increased. As in any other regional-scale assessment, there is a large degree of uncertainty associated with this study. However, this should not discourage risk managers from utilizing this assessment. We acknowledged the high degree of uncertainties associated with this risk assessment and, therefore, only broad risk categories: high, medium, low, and no risk were concluded from this study. Uncertainty for the risk assessment can be greatly reduced if additional stressors, habitats, and exposure pathways data are available, especially in regions 3, 5, and 6. Management decisions within the Squalicum Creek watershed are currently made on a case-by-case basis that only addresses the ecological effects of individual parcel development. These assessments ignore any effects resulting from the interactions between sources and receptors from different parcels. Various uncertainties are associated with these decisions due to the potential of combined effects of sources from separate parcels. This assessment can integrate the effects of multiple individual decisions and, therefore, risk managers can make decisions that would minimize the adverse impacts to the watershed. It also allows the risk managers to prioritize the importance to the watershed of the parcels according to their rankings. This is particularly important for the Squalicum Creek watershed because of its unique position in which the port, the city, and the county all have jurisdiction over the area. The rankings can enable these authorities to have a common set of priorities for the parcels and, therefore, make decisions that would minimize the adverse impacts to the watershed.

This assessment also serves as a framework to organize the existing data and point out where data are lacking. The uncertainty analysis can help the risk managers identify areas of research that could minimize the most model uncertainties. Another benefit of the assessment to the management process is that it enables the stakeholders to estimate the effects of different management options to the watershed. The RRM was designed in a format that is easy to use and understand, so stakeholders can utilize the results of the model to conduct cost–benefit analyses. With acknowledgment of uncertainties and suggestions for future improvements, this assessment can provide risk managers with a comprehensive tool to aid in future decisions in the Squalicum Creek watershed.

CONCLUSION

The application of the RRM to the Squalicum Creek watershed risk assessment was successful. A few modifications to the model were made due to the addition of risk components. The sensitivity analysis results demonstrate that the model produces nonbiased results. We also found that the model is more sensitive to combined effects from variation of multiple components and that the model is less sensitive to variations to filters. The risk assessment results indicate that residential landuse contributed most risk to the watershed. The results also show that the coldwater habitat and wetland endpoint are most at risk in the study area. The salmonids endpoint was found to be at relatively low risk because of the limited access to habitats due to physical barriers. We also found that there is a strong risk gradient in the study area in which risk increases from upstream regions to downstream regions. The risk assessment shows that lowering the number of stressors and isolating the habitats can reduce more risk than increasing the number of habitats. Uncertainties for this study mostly come from the lack of data regarding the sources and habitats, especially in regions 3, 5, and 6. Despite all the uncertainties for this study, risk assessment can have application to the management of the Squalicum Creek watershed. By acknowledging the uncertainties for this risk assessment, the authorities in the watershed can better understand where and what data are needed most in the watershed. The authorities can also use the risk assessment result to set a common priority for the parcels. This can enable them to utilize the best existing data to make decisions that would minimize the adverse impacts to the watershed.

PART II: RISK PREDICTION TO MANAGEMENT OPTIONS IN THE SQUALICUM CREEK WATERSHED USING THE RELATIVE RISK MODEL ECOLOGICAL RISK ASSESSMENT

Introduction

Management of regional-scale ecosystems is challenging to many risk managers due to the limitations in current resource management methodologies. The objective of most resource managers is to identify the management option that would produce results closest to the management goal. Resource managers need to define the management goal, determine the current state of the ecosystem, and predict the future conditions of the system and the effects of management decisions on the system.

One of the difficulties in resource management is defining the management goal. Many attempts by risk managers are unsuccessful due to the use of vague terms such as "ecological health," "ecological integrity," and "recovery." These terms have many definitions and, therefore, managers do not have a uniform perception of them. Application of these terms to the management process is also difficult because they are not parameters that can be measured directly (Landis and McLaughlin 2000a, 2000b). All of these terms require the use of surrogate variables that might lead to uncertainties. For example, the use of a single species or biological index number to indicate an ecosystem is insufficient because these parameters oversimplify the complexity of the system (Landis and McLaughlin 2000a).The definitions of these vague terms could also lead to impractical goals. For example, many managers define their management goal as recovery to a preexisting condition prior to human impact, or recovery to a condition of a reference site. These are often unachievable goals because ecosystems inherit ecological effects from the past (Landis et al. 1993a; 1993b) and, therefore, returning to those conditions might not be possible. Also, reference sites as strictly defined do not exist since no two sites are identical. Recovery to the condition of a reference site is irrational because the two sites were not the same and never will be the same due to inherent differences.

Another difficulty in resource management is the limited predictability of management methods to future conditions. Most of the current management techniques only allow the risk managers to describe the current ecosystem condition; very few techniques allow for prediction of future ecosystem changes. One of the most commonly used is adaptive management. It is a technique accepted by many risk managers and involves experimenting with the response of the ecosystem to human behavior changes in the systems (Lee 1999). Although the concept of adaptive management is widely accepted, there are currently insufficient case studies to prove the utility of this management method. Very few cases have utilized this method, many of which have not obtained results due to the slow response of the ecosystems. Other cases do not meet all the requirements of the management method. One of the drawbacks of adaptive management is that it is often costly and time consuming because a large amount of data is needed to test each hypothesis. The methodology does not allow the managers to compare the effects of the decision to those of alternative decision options. The methodology is also difficult to apply to regional-scale ecosystems due to system dynamics (Lee 1999).

Table 10.8   Percentage Risk Change for Each Endpoint in Each Risk Region for Decision Option 1 (numbers represent percentages and negative numbers indicate reduced risk to endpoint)

Risk Region	Nonmigratory Cold- Water Fish	Life Cycle Opportunities for Salmonids	Flood Control	Wildlife Species	Recreational Uses	Wetland
1	0	0	0	0	0	0
2	0	0	0	0	0	0
3	−8.6	82.8	−12.0	−4.3	−10.2	−6.9
4	−9.2	81.5	0	−6.1	−10.9	−7.6
5	0	0	0	0	0	0
6	0	*	0	0	0	0

* Changed from no risk to approximately the risk of the current condition of the salmonids in region 4.

Subsequently, we have chosen an alternative approach to risk management, the RRM. The RRM was successfully applied to a number of other risk assessments of ecosystems with various scales, stressors, and endpoints (Landis and Wiegers 1997; Obery and Landis 2002; Walker et al. 2001). The study conducted by Thomas et al. (2001) further confirmed the utility of the RRM as a tool to analyze alternative decisions. The RRM follows the management concept established by Landis and McLaughlin (2000b) in which the management goal is defined by the movement of endpoints in relation to the stakeholders' set limits over time. The RRM incorporates stakeholders' values and defines the management goal using measurable endpoints. The model allows the risk managers to determine the current condition of the endpoints, predict future conditions, and predict the effects of management decision options on these endpoints. Unlike many other management techniques, the RRM requires minimal cost and time because the model utilizes existing data. It also requires relatively few data to confirm the model results. The model can be applied to various-scale ecosystems, and it can incorporate the effects of multiple (biological, chemical, and physical) stressors to multiple endpoints. The model is also set up in a way that is easy to use and understand by the managers, and the model is very flexible so it can be modified easily in response to changes in model variables.

The objective of this study was to test the utility of the RRM as a tool for analyzing decision options. Among the studies that used the RRM method, we chose to analyze the Squalicum Creek watershed RRM because the watershed is developing rapidly and the level of concern has increased considerably in recent years. The Squalicum Creek watershed is also the smallest ecosystem that has been analyzed using the RRM. This watershed serves as a good example for a typical urbanized watershed in the Pacific Northwest, and can demonstrate the potential of the application of the RRM to similar ecosystems in the area. In this study, we first developed a list of possible decision options for the Squalicum Creek watershed. We then determined the relative risks to the watershed due to various decision options. Last, we compared these risk predictions to the risk results of the current Squalicum Creek watershed as determined in Part I of this project.

METHODS

The methodology of using the RRM to predict the impact of decision options to an ecosystem varies slightly for individual ecosystems due to differences in the RRM. In all cases, the researchers need to:

- Conduct studies to include an EcoRA for the study area using the RRM to determine the current condition of the area
- Define a list of decision options
- Recalculate the input data such as habitat and sources of stressors data according to the decision options
- Rerank the new input data from step 3 using the ranking criteria in the RRM in step 1
- Alter the exposure pathways in the model according to the decision options and enter the subsequent data from step 4 into the model to generate risk scores

- Calculate the change in risk scores from step 1 results to step 5 results
- Analyze and compare the results from step 6 for different decision options

RISK ASSESSMENT

Step 1 is included in Part I of this project.

LIST OF DECISION OPTIONS

We selected the following six decision options based on proposed action in the study area and stakeholders' values. These six options are not real decision options that the resource managers are currently undertaking; instead, they are potential decision options that are likely to be achievable based on the current condition and developmental trends in the Squalicum Creek watershed.

Option 1: Convert the Impassable Culverts to Passable Culverts

The stakeholders have expressed their concern about the impact of impassable culverts on migratory aquatic species such as the salmonids. Many stakeholders have a misconception that the removal of impassable culverts would decrease risks to species affected by these culverts. We included this decision option to help the stakeholders better understand and predict the possible risk changes resulting from this option.

Option 2: Increase 25 and 50%, Respectively, of Forested Area in Agricultural Land Riparian Corridor

Since October 1998, the Conservation Reserve Enhancement Program (CREP) has become available to landowners in Whatcom County. The intention of the program is to help restore the riparian buffers along salmon- and steelhead-supporting streams to help improve the habitats of these species. It is a voluntary program in which landowners of agricultural land that meets specific requirements are qualified to participate in the program. Some of these requirements include cropping history, stream designations, and riparian buffer width in the property. Program participants are required to stop any agricultural activities in the designated CREP area on their property for a period of 10 to 15 years. They are also required to plant and maintain native vegetations in those designated areas. In return, the Farm Services Agency and the Conservation Commission pay the program participants an annual rent for their property and also cover the other expenses needed for the habitat restoration process. Since CREP became available in Whatcom County, various landowners have signed up for the program. With the growing concern for the salmon habitats in the county, we assumed there would be a continuous growth in the number of CREP participants in the Squalicum Creek watershed. In this decision option, we assumed a 25 to 50% CREP participation for all the agricultural landowners who are within the Squalicum Creek watershed, including sections of the 60.96-m riparian buffer. This leads to a 25 to 50% increase of forested area in agricultural land riparian corridor in the study area. We included this decision option so resource managers could better estimate the efficiency of this decision option in the protection of aquatic species.

Option 3: Eliminate Forestry Activities

With the continuous decrease of habitats in the Squalicum Creek watershed and the stakeholders' increasing concern of risks to wildlife habitats, it would not be surprising for the resource managers to eliminate forestry activities in the watershed in the future.

Option 4: No Action — Resulting in 100% Development in Undeveloped and Forested Land in Urban Growth Area

This is a very probable decision option in comparison to all other options. If no action is taken in the near future, it would result in approximately 100% development in the undeveloped and the privately owned forested land in the urban growth area in the study area. We included this no-action option so resource managers could compare the impact of other options to this one and could pick the decision option that best meets their management goal.

Option 5: Divert Storm Runoff from Industrial and Commercial Areas to Treatment Facilities

With the increasing use of retention ponds in industrial and commercial landuses in the Squalicum Creek watershed, it would become more feasible and possible in the future to divert the storm runoff from these landuses to treatment facilities. We included this option so resource managers could estimate the effect of this option to the Squalicum Creek watershed. For this decision option, we assumed a 100% efficiency of the treatment facilities. In the RRM, we removed the increased runoff and increased chemical stressors from all industrial and commercial landuses.

Option 6: Eliminate Mining Activities

Mining activities produce many stressors that might increase risks to the ecosystem. Mining activities are currently found only in regions 1 and 2 of the study area. The RRM results showed that these areas are currently at the highest risk (see Part I of this project). We included this decision option to see if eliminating the mining activities could change the overall risk rank (high, medium, or low) for risk regions 1 and 2.

UNCERTAINTY ANALYSIS

There are two major sources of uncertainties to this study. The first source is the uncertainties from the risk assessment performed in Part I of this project. The second source is the uncertainties associated with the lack of knowledge regarding the data input into the RRM for the decision options. The following is a list of general uncertainties identified in the risk assessment for the current condition of the watershed (see Part I of this project):

- Stochastic nature of ecological and human systems, such as the random variations of endpoint response to stressors, leading to nonreducible uncertainties
- Human error
- Lack of information regarding the completeness of exposure pathways, leading to uncertainties in the conceptual model
- Insufficient temporal data, leading to unknown seasonal patterns
- Lack of coldwater habitat data, leading to the use of the same risk rank (high rank) for coldwater habitat for all risk regions
- Insufficient data for stressors, habitats, and possible endpoint locations in the Squalicum Creek watershed, especially in regions 3, 5, and 6, leading to uncertainties regarding to the use of landuse data

The following is a list of uncertainties associated with the lack of knowledge regarding the data input into the RRM to predict the effects of decision options to the study area:

- Due to limited knowledge, the input data we used to predict the effects of decision options did not account for seasonal patterns, which might lead to incorrect predictions.
- The relative position of the endpoints in relation to the stakeholders' set limit for the endpoints cannot be concluded from this study because there are currently insufficient data to confirm the position of the endpoints.
- Migratory properties of endpoints within each risk region are not addressed in this study.
- The risk prediction for the decision options does not account for sensitive habitats; each habitat is treated with equal weight in the study.
- The effects of the amount of storm runoff to the watershed from industrial and commercial landuses are not accounted for due to the limited data regarding the stressor groups.
- The amount of development in the study area for decision option 4 can only be estimated because of unknown actions of the landowners.

As a part of the uncertainty analysis, we performed a sensitivity analysis to address the degree of variation of the predicted risks for the different decision options. We evaluated decision option 4 because it is the most probable decision option and can be tested for a range of input data. We included different percentages of development for the no-action decision option ranging from 10 to 100% at 10% intervals.

RESULTS

Risk Changes to Option 1:
Convert the Impassable Culverts to Passable Culverts

The RRM predicts that this decision option would decrease risk in region 3 and increase risk in regions 4 and 6 (Figure 10.6). This is a result of a decreased risk to

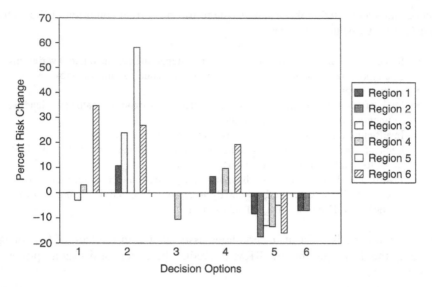

Figure 10.6   Percentage risk change in each risk region for each decision option.

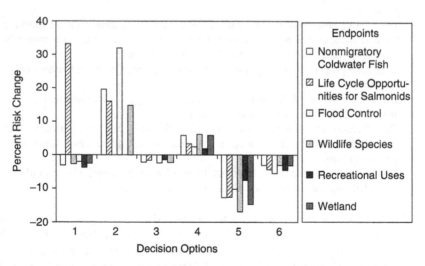

Figure 10.7   Percentage risk change to each endpoint for each decision option.

all but the salmonids endpoint in regions 3 and 4 and a relatively large increase of
risk to salmonids in region 6. In region 3, the increased risk to the salmonids endpoint
is outweighed by the decreased risk to the other endpoints. As a result, risk to region
3 decreased, whereas in regions 4 and 6, the increased risk to salmonids endpoints
outweighed the decreased risk to all other endpoints combined. Especially in region
6, the risk to salmonids changed from no risk to approximately the same risk to
salmonids in the current condition of region 4 (Table 10.1). As a result, risk to
regions 4 and 6 increased. The RRM also predicts a decreased risk in all but the

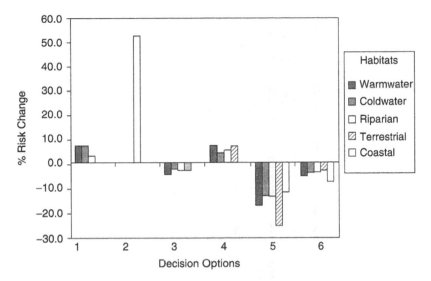

Figure 10.8 Percentage risk change to each habitat for each decision option.

Table 10.9 Percentage Risk Change from Each Stressor Group for Each Decision Option (numbers represent percentages and negative numbers indicate reduced risk to endpoint)

Decision Option	IR	IC	SFA	IN	AF	AW	IS	ITS
CC	5.1	5.4	−3.0	6.1	4.2	−3.0	5.0	5.0
IFA	12.1	12.8	11.0	15.4	10.0	11.0	0	42.1
EF	−1.3	−1.7	−1.1	−2.6	−1.3	−1.1	−4.2	0
NA	4.7	4.7	3.6	7.1	4.3	3.6	0	13.5
DSR	−39.2	−38.1	0	0	0	0	0	0
EM	−3.9	−3.6	−3.9	0	−4.0	−3.9	−12.9	0

CC	Convert Culverts	NA	No Action
IFA	Increase Forested Area	DSR	Divert Storm Runoff
EF	Eliminate Forestry	EM	Eliminate Mining
IR	Increased Runoff	AF	Altered Forest
IC	Increased Chemicals	AW	Altered Wetland
SFA	Stream Flow Alteration	IS	Increased Sediments
IN	Increased Nutrients	ITS	Introduced Terrestrial Species

salmonids endpoint (Figure 10.7), an increased risk in coldwater habitat, warmwater habitat, and riparian habitat (Figure 10.8) and an increased risk from all but altered stream flow and altered wetland stressor groups (Table 10.9). Other than the elimination of risk from physical barrier construction, this decision option is predicted to increase risk from all landuses except mining (Table 10.10).

Table 10.10 Percentage Risk Change from Each Stressor Source for Each Decision Option (numbers represent percentages and negative numbers indicate reduced risk to endpoint)

Decision Option	Agricultural	Residential	Light Industrial	Heavy Industrial	Mining	Chemical Industries	Commercial	Park	Transportation	Forestry Activities	Dams and Culverts
CC	5.9	6.5	2.0	3.7	0	13.2	5.1	3.1	3.6	13.2	-100.0
IFA	19.6	15.1	15.3	8.6	4.4	11.5	12.0	6.6	10.3	0	12.6
EF	0	0	0	0	0	0	0	0	0	-100.0	0
NA	0	19.5	22.4	0	0	0	0	0	0	0	0
DSR	0	0	-30.8	-31.0	0	-39.9	-36.0	0	0	0	0
EM	0	0	0	0	-100.0	0	0	0	0	0	0

CC   Convert Culverts   NA   No Action

IFA   Increase Forested Area   DSR   Divert Storm Runoff

EF   Eliminate Forestry   EM   Eliminate Mining

Risk Changes to Option 2: Increase 25 and 50% of Forested Area in Agricultural Land Riparian Corridor

A 25 and 50% increase of forested area in agricultural land is predicted to have the same impact on the Squalicum Creek watershed. This decision option is predicted to increase risks to regions 2, 3, 5, and 6, with the greatest increase designated to region 5 (Figure 10.6). It is also predicted to increase risk to all biotic endpoints, especially to the terrestrial foreign species (Figure 10.7). The RRM also predicts this decision option to increase risk in riparian habitat (Figure 10.8), increase risk from all stressor groups except increased sediment (Table 10.9), and increase risk from all landuses except forestry activities (Table 10.10).

Risk Changes to Option 3: Eliminate Forestry Activites

This decision option is predicted to reduce risk in region 4, the only risk region associated with forestry activities (Figure 10.6). It is also predicted to reduce risk to all but the flood control endpoint (Figure 10.7), reduce risk occurring in all but coastal habitat (Figure 10.8), and reduce risk from all except the introduced terrestrial foreign species stressor group (Table 10.9). Other than eliminating forestry activities in the Squalicum Creek watershed, risk from all other landuses is predicted to remain unchanged (Table 10.10).

Risk Changes to Option 4: No Action — Resulting a 100% Development in Undeveloped and Forested Land in Urban Growth Area

If no action is taken in the Squalicum Creek watershed, risks to regions 2, 4, and 6 are predicted to increase, with the greatest increase occurring in region 6 (Figure 10.6). It is also predicted that this decision option would lead to an increased risk to all endpoints (Figure 10.7), an increased risk in all except coastal habitat (Figure 10.8), an increased risk from all except the sediment stressor group (Table 10.9), and an increased risk from residential and light industrial landuses (Table 10.10).

Risk Changes to Option 5: Divert Storm Runoff from Industrial and Commercial to Treatment Facilities

This decision option is predicted to decrease risk to all risk regions (Figure 10.6), all endpoints (Figure 10.7), and all habitats, especially to the terrestrial habitat (Figure 10.8). It is also predicted to reduce risk from increased runoff and increased chemical stressor groups (Table 10.9), and reduce risk from all industrial and commercial landuses (Table 10.10).

Risk Changes to Option 6: Eliminate Mining Activities

The elimination of mining activities in the Squalicum Creek watershed is predicted to reduce risk to regions 1 and 2, the only regions affected by mining activities (Figure 10.6). This decision option is also predicted to reduce risk to all endpoints

(Figure 10.7) and all habitats (Figure 10.8) and reduce risks from all except increased nutrients and introduced terrestrial foreign species stressor groups (Table 10.9). Other than eliminating risk from mining landuse, risk from other landuses remains unchanged (Table 10.10).

Sensitivity Analysis Results

The results remained the same for development ranging from 50 to 100%. The risk rank for light industrial landuse in region 6 changed from a 4 as in the 50 to 100% range to a risk rank of 2 in the 10 to 40% range. The percentage risk change for region 6 also dropped from 19.7% as in the 50 to 100% range to 9.8% in the 10 to 40% range. There were relatively few risk changes with the range of development for option 4 because most of the percentages for development in the original RRM were located close to the breakpoint. In most cases, a slight increase of the percentage of development would result in an increase of risk rank by 2, while a relatively large amount of development (greater than 100%) is required to increase the risk rank by more than 2. The risk rank for light industrial landuse in region 6 did not change with the range of development because there was no light industrial landuse in region 6 in the original RRM. In summary, the sensitivity analysis shows that the risk component risk rank is more sensitive to small changes when the location of the risk component is closer to the ranking breakpoint.

DISCUSSION

The RRM results demonstrated that in most cases, reduction of stressor sources and stressors could reduce the most risks in the watershed as in the case of options 3, 5, and 6. The RRM predicts that option 1, however, would decrease risk to all but the salmonids endpoint. The removal of culverts is predicted to reduce risk to the nonmigratory species because this removal would lead to a reduction of the amount of stressors resulting from these culverts. However, removal of culverts would allow the salmonids to inhabit new habitats upstream of the current culvert locations. Introduction of salmonids to these new habitats could increase their exposure to greater amounts of stressors, and risk to salmonids would increase, as predicted by the RRM. The RRM predicts that if no action were to be taken in the study area, i.e., option 4, there would be an increase of risk to regions 2, 4, and 6, which are located in the urban growth area.

The model also predicts that creating new habitats without reducing the amount of stressors as in option 2 would lead to an increase of risks to the endpoints. This is not necessarily true, because by increasing the area of habitat in a risk region while having the same amount of stressors, the effects from stressors could be diluted. As a result, risk to endpoints in these habitats would decrease. The RRM predictions did not reflect this decrease due to the model limitation associated with one of the assumptions of the model. The model assumes that exposure increases as one or more of the following risk components increase in a region: stressor source density, the number of stressors, and habitat density. This assumption does not take into account the dilution property of increased habitat density toward the amount of sources and stressors in a region. As a result, endpoint risk predictions from the model might be overestimated in regions with increased habitat density. Risk managers should be aware that this model might falsely predict the risk results in cases where the habitat density is increased.

Model limitations specific to the application of the RRM to the Squalicum Creek watershed, however, should not discourage risk managers from utilizing the RRM as a risk predictive tool. In most cases, the RRM allows risk managers to estimate the effects of decision options on the study area and, therefore, allows them to choose the decision option that would produce the result that is closest to their management goal. The results of the RRM in predicting the effects of the decision options suggest that the most effective way of decreasing risk in the study area is to disrupt the exposure pathway or to decrease the amount of sources and stressors entering into the habitat. However, this should be planned carefully, because, as in the case of the removal of the culverts, some endpoints might be introduced into new habitats, leading to an increased risk to the endpoints. There is often a misconception that the removal or reduction of stressors would reduce risk. The RRM is especially useful for risk managers in these cases, because it allows them to better understand the possible effects of this type of decision option so they can modify their decision options. For example, for option 4, risk managers could expand the option by including a reduction of risks in the upstream regions of the culverts. The modified option could produce a result that is closer to their goal. This process can be repeated in a relatively short time and at a low cost to produce a final decision option that would best fit the risk managers' values.

CONCLUSIONS

This study shows the utility of the RRM as a predictive model for the effects of decision options on the study area. The RRM is a good tool for predicting the effects for all decision options except when the decision option involves increasing or creating habitats. The model prediction demonstrated a clear movement of the endpoints in response to the decision options. The prediction enables risk managers to select the decision option that would produce the closest result to their management goal with a relatively small amount of data collection and time. This study, however, also shows that there are a few limitations and uncertainties associated with the use of the RRM as a predictive tool. A few of these limitations include the inability of the model to account for seasonal patterns and migratory patterns of endpoints within a risk region and the potential uncertainty associated with sensitive habitats. This RRM for the Squalicum Creek watershed can be greatly improved if studies were to be conducted to confirm the endpoint positions. Modifications to the model to incorporate the model limitations identified in this study, such as the inability to account for the dilution property of increased habitats toward a fixed amount of stressors, could also improve the accuracy of the model significantly.

REFERENCES

Hoerauf, E. 1999. Western Washington University, personal communication.

Landis, W.G. and McLaughlin, J.F. 2000a. Design criteria and derivation of indicators for ecological position, direction, and risk, *Environ. Toxicol. Chem.*, 19, 1059.

Landis, W.G. and McLaughlin, J.F. 2000b. If not recovery, then what?, in *Environmental Toxicology and Risk Assessment: Science, Policy and Standardization – Implications for Environmental Decisions*, Tenth volume, Greenburg, R.N. et al., Eds., ASTM STP 1403. B.M., American Society for Testing and Materials, West Conshohocken, PA.

Landis, W.G., Moore, R.J., and Norton, S. 1998. Ecological risk assessment: looking in, looking out, in *Pollution Risk Assessment and Management*, Ed. P. E. T., Douben, J.W., Sons Ltd., Chichester..

Landis, W.G. and Wiegers, J. 1997. Design considerations and a suggested approach for regional and comparative ecological risk assessment, *Hum. Ecol. Risk Assess.*, 3, 287.

Landis, W.G. et al. 1993a. Multivariate analyses of the impacts of the turbine fuel Jet-A using a microcosm toxicity test, *Environ. Sci.*, 2, 113.

Landis, W.G. et al. 1993b. Multivariate analysis of the impacts of the turbine fuel JP-4 in a microcosm toxicity test with implications for the evaluation of ecosystem dynamics and risk assessment, *Ecotoxicology*, 2, 271.

Lee, K.N. 1999. Appraising adaptive management, *Conserv. Ecol.*, 3, 3. http://www.eco-secol.org/vol3/iss2/art3.

Obery, A.M. and Landis, W.G. 2002. A regional multiple stressor risk assessment of the Codorus Creek watershed applying the relative risk model, *Hum. Ecol. Risk Assess.*, 8, 405.

Thomas, J.F. et al. 2001. Confirmation of a Relative Risk Model ecological risk assessment of multiple stressors using multivariate statistics, Western Washington University, Bellingham, WA.

USEPA (U.S. Environmental Protection Agency). 1998a. Comparative ecological risk: using the proximity of potential and actual stressors to resources as a tool to screen geographical areas for management decisions comparative geographical risk assessment, in EPA Region 10 Development of a Prototype, Duncan, B. et al., Eds., Office of Environmental Assessment.

USEPA (U.S. Environmental Protection Agency). 1998b. Guidelines for Ecological Risk Assessments, Risk Assessment Forum, Washington, D.C., EPA/630/R-95/002F.

USEPA (U.S. Environmental Protection Agency). 1998c. Better Assessment Science Integrating Point and Nonpoint Sources (BASINS), Office of Water, Version 2. USGS (U.S. Geological Survey). 2000. http://duff.geology.washington.edu/data/raster/tenmeter/byquad/victoria/index.html.

USGS (U.S. Geological Survey). 2000. 10-Meter DEM Files in Victoria. http://duff.geology.washington.edu/data/raster/tenmeter/byquad/victoria/index.html.2000.

Walker, R., Landis, W., and Brown, P. 2001. Developing a regional ecological risk assessment: a case study of a Tasmanian agricultural catchment, *Hum. Ecol. Risk Assess.*, 7, 417.

Whatcom County Council. 2000. Whatcom County Code, Whatcom County Council, WA.

Whatcom County Assessors Office. 2000. Whatcom County Land Use Codes, Whatcom County Assessors Office, WA. http://www.mrsc.org/cgi_bin/om_cgi_exe?&infobase=whatcom.nfo&softpage=Browse_

Whatcom Conservation District. 2000. Fish Presence Mapping Project for WRIA 1, Whatcom Conservation District, WA.

Wiegers, J.K. et al. 1998. A regional multiple-stressor rank-based ecological risk assessment for the fjord of Port Valdez, Alaska, *Hum. Ecol. Risk Assess.*, 4, 1125.

Appendix A

LIST OF EQUATIONS FOR WIEGERS ET AL.'S (1998) STUDY

$$RS = S_{ij} \times H_{ik} \times W_{jk}$$ Equation (10.1)

where:
RS = final risk scores
i = the subarea series
j = the source series (1...8)
k = the habitat series (1...8)
S_{ij} = rank chosen for the sources between subareas
H_{ik} = rank chosen for the habitats between subareas
W_{jk} = weighting factor established by the filters

$$RS_{Source} = \sum (S_{ij} \times H_{ik} \times W_{jk}) \text{ for } j = 1...8 \quad \text{Equation (10.2)}$$

$$RS_{Habitat} = \sum (S_{ij} \times H_{ik} \times W_{jk}) \text{ for } k = 1...8 \quad \text{Equation (10.3)}$$

LIST OF EQUATIONS FOR THIS STUDY

$$R_{Endpoint_n} = \sum (S_{ij} \times H_{il} \times C_{jk} \times BS_{kl} \times BE_{ln}) \text{ for } n = 1...3 \quad \text{Equation (10.4)}$$

where:
R = risk score
i = risk region series (region 1...region 6)
j = source series (agricultural landuse...stream barrier construction)
k = stressor series (increase runoff...introduced terrestrial species)
l = habitat series (coastal...terrestrial)
n = all biotic endpoints except salmonids endpoint (nonmigratory coldwater fish population...wetland)
C = sources of stressors filter
BS = biotic stressor filter
BE = biotic effect filter
S_{ij} = rank chosen for the sources between regions
H_{il} = rank chosen for the habitats between regions

$$R_{Endpoint_o} = \sum (S_{ij} \times H_{il} \times C_{jk} \times BS_{kl} \times BE_{lo} \times T_{lt} \qquad \text{Equation (10.5)}$$

where:
o = salmonids endpoint
T = habitat filter

$$R_{Endpoint_n} = \sum (S_{ij} \times E_{ip} \times C_{jk} \times AS_{kp} \times AE_{kq} \text{ for } q = 1...2 \qquad \text{Equation (10.6)}$$

where:
p = possible endpoint location series (recreational uses endpoint location...flood control endpoint location)
q = abiotic endpoints (recreational uses...flood control)
AS = abiotic stressor filter
AE = abiotic effect filter
E_{ip} = rank chosen for the possible endpoint locations between regions

$$R_{Stressor} = \sum (S_{ij} \times H_{il} \times C_{jk} \times BS_{kl} \times BE_{ln} + S_{ij} \times H_{il} \times C_{jk} \times BS_{kl} \times BE_{lo} \times T_{lo}$$
$$+ S_{ij} \times E_{ip} \times C_{jk} \times AS_{kp} \times AE_{kq})$$

for $k = 1...8$ Equation (10.7)

$$R_{Source} = \sum (S_{ij} \times H_{il} \times C_{jk} \times BS_{kl} \times BE_{ln} + S_{ij} \times H_{il} \times C_{jk} \times BS_{kl} \times BE_{lo} \times T_{lo}$$
$$+ S_{ij} \times E_{ip} \times C_{jk} \times AS_{kp} \times AE_{kq})$$

for $j = 1...11$ Equation (10.8)

$$R_{Stressor} = \sum (S_{ij} \times H_{il} \times C_{jk} \times BS_{kl} \times BE_{ln} + S_{ij} \times H_{il} \times C_{jk} \times BS_{kl} \times BE_{lo} \times$$

for $l = 1...5$ Equation (10.9)

$$R_{Region} = \sum (S_{ij} \times H_{il} \times C_{jk} \times BS_{kl} \times BE_{ln} + S_{ij} \times H_{il} \times C_{jk} \times BS_{kl} \times BE_{lo} \times$$
$$+ S_{ij} \times E_{ip} \times C_{jk} \times AS_{kp} \times AE_{kq})$$

for $i = 1...6$ Equation (10.10)

CHAPTER 11

The Use of Regional Risk Assessment in the Management of the Marine Resources in the Cherry Point Region of Northwest Washington

April J. Markiewicz

CONTENTS

1-56670-655-6/04/$0.00+$1.50
© 2004 by CRC Press LLC

MANAGEMENT OF THE MARINE RESOURCES IN THE CHERRY POINT REGION OF NORTHWEST WASHINGTON

The Washington Department of Natural Resources (WDNR) manages the aquatic lands of Washington state "for current and future citizens of the state to sustain long-term ecosystem and economic viability" (WDNR 2001). This joint mission to protect the natural resources of the state while at the same time generating income by harvesting those resources creates a framework in which difficult management decisions must be made. The purpose of this regional risk assessment using the relative risk model (RRM) was to provide estimates of the historical, current, and future relative risks that potentially affected or will affect the Cherry Point (CP) herring stock, as well as the surrounding region from nearby anthropogenic sources to aid WDNR in these management decisions. Moreover, alternative assessment endpoints were also identified in the CP region that would provide additional information to WDNR regarding the chemical, toxicological, organismic, and trophic interactions specific to the CP area. These additional assessment endpoints in combination with the results of the CP regional-scale ecological risk assessment will help to ensure responsible management of the ecological resources by WDNR in the CP region.

INTRODUCTION

The WDNR is responsible for the protection of public resources and management of all state-owned land, as well as aquatic lands that include tidelands and lands beneath the waters of Puget Sound. As stewards they are charged with managing these resources and lands to provide benefits for current and future citizens of the state. Those benefits include providing public access for recreation, supporting business activities that depend on access to water, and ensuring protection of the state's natural resources. The decisions that WDNR makes involve:

- Leases, easements, rights-of-way
- Aquatic reserve management decisions
- Restoration projects
- Public access decisions
- Actions to prevent "takings" under the Endangered Species Act

The specific decisions that WDNR has to make with regard to the aquatic lands along the reach at the Cherry Point, Washington region (Figure 11.1) include the following:

- Alcoa-Intalco Aluminum Company pier and outfall lease renewal
- Alcoa-Intalco Aluminum Company stormwater outfall (perpetual easement)
- Lummi Indian Business Council waste outfall at Neptune Beach
- Pacific International Terminal proposal for dock expansion at the BP Amoco pier
- Williams Pipeline request to run their pipeline through Cherry Point out into the Strait of Georgia
- Whatcom County-owned waste outfall
- ConocoPhillips (Tosco) Ferndale Refinery pier and outfall lease renewal
- BP Amoco lease compliance monitoring

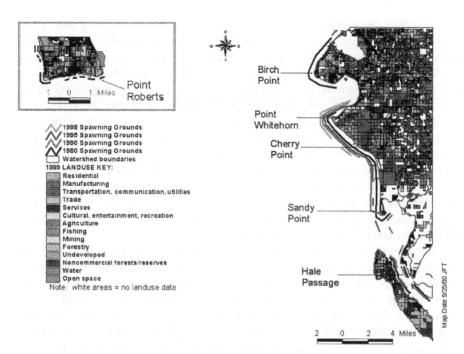

Figure 11.1   CP region herring spawning grounds: 1980, 1990, 1995, and 1998. (Prepared by J. Thomas, data from Gonyea et al. 1982; Stick, 1990, Lemberg et al. 1997; Whatcom County Planning and Development Services, 2001.) (See color insert following page 178.)

- Management of CP aquatic reserve — determine what actions are compatible with goals for the region
- Evaluate ongoing and future restoration and mitigation projects in the region
- Evaluate impacts of recreational uses (like clam digging) on the area
- Determine how to deal with abandoned bulkheads/armored structure
- Determine how to deal with outfalls not under WDNR lease

In response to growing concerns that their leasing and other management decisions at Cherry Point may have adversely impacted the Pacific herring stock that spawn there annually, WDNR contracted to have a screening-level ecological risk assessment (EcoRA) conducted in 1999 (EVS Consultants, LTD 1999) and a more extensive EcoRA conducted in 2000 (Landis et al. 2000). The focus of this chapter is to introduce the background information, as well as the decision-making needs and management goals that helped to direct the more extensive CP EcoRA conducted in 2000 that are described in the following chapter.

BACKGROUND

Cherry Point

Cherry Point is located on the northwest coast of Washington state and is considered one of the last undeveloped deepwater ports in the United States. Until the mid-1900s the land was relatively undeveloped with some forestry and agricultural farming. The land, shoreline, and nearshore areas along the CP reach extending from Point Whitehorn to Sandy Point were reknowned for the diverse species that resided and utilized the natural resources in the area. Moreover, the reach also served as the primary spawning site for a population of Pacific herring, supporting a successful herring sport and commercial bait and roe fishery over the years.

In the late 1940s the deep water close to shore and the steep gradient of the intertidal zone along the Cherry Point reach were recognized for their potential as a deepwater port. Construction began shortly afterward, and by the 1970s three refineries and their associated piers had been constructed and were in full operation. The refinery located just to the north of Cherry Point proper is currently owned by the British Petroleum (BP) Amoco; just to the south of Cherry Point is the Alcoa-Intalco Aluminum Company (Intalco) refinery, and to the south of it is the Conoco-Phillips (formerly Tosco) Ferndale refinery.

Although there have been concerns over the years that the operations of the refineries and their shipping piers could potentially cause adverse impacts to the environment, recent studies have indicated that species diversity along the reach has remained high (Kyte 2000; 2001). However, the CP herring stock that migrate back to the region each spring to spawn have been returning in fewer numbers since the mid 1970s. Moreover, the age class structure of the population has been altered with only younger (less than 4 years old) fish comprising the bulk of the stock.

CP Herring Stock

Pacific herring are central to sustaining the natural structure, function, and biodiversity of the Puget Sound marine ecosystem due to their trophic importance as a high quality, abundant food source. Piscivorous marine fish including salmonids, Pacific hake, and Pacific cod, as well as marine mammals including harbor seals, sea lions, whale, and porpoises, and coastal bird populations such as sea gulls, cormorants, murrelets, and great blue herons forage extensively on the eggs, larvae, juvenile and adult life stages of Pacific herring. The herring are also a valued economic resource and have been harvested for bait fisheries, fish meal production, and the overseas roe market.

The CP herring stock is one of 18 Pacific herring stocks in the state, and in 1973 was the largest herring stock in Puget Sound (Bargmann 1998; Penttila and Burton 1986, Penttila et al. 1986). After a recorded high abundance of almost 15,000 tons of herring at Cherry Point and in the surrounding area in 1973, the number of CP herring steadily began to decline (Bargmann 1998; Lemberg et al. 1997; Penttila and Burton 1986, Penttila et al. 1986). Attempts to slow or prevent further declines that included curtailing fishing quotas and stopping harvests entirely in 1981, 1983–1986, and from 1997 to the present were ineffective (Figure 11.2). As of 1998, the stock size was estimated to be at 1322 tons, less than one tenth its population level in 1973. Concurrently there has also been a reduction in the age structure of the population so that by 1998 almost 80% of the population was comprised of 2-year olds. Six- to nine-year age classes of CP herring were almost completely absent,

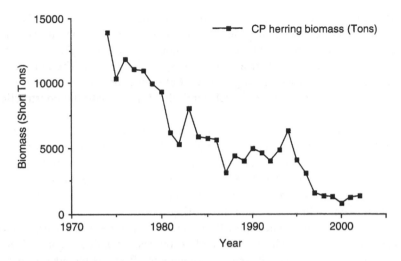

Figure 11.2 CP herring stock biomass from 1974 to 2002.

with relatively few individuals found in the 4- and 5-year age classes (Bargmann 1998; EVS 1999; Lemberg et al. 1997). By 2000 the CP herring stock size reached a population low of approximately 808 tons.

The severity of the CP herring stock decline was considered unique due to the sheer numbers and biomass of herring that disappeared from the region compared to the other Puget Sound herring stocks (EVS 1999). As a result, state and local agencies, as well as area residents, Lummi and Nooksack Indian tribes, environmental groups, and sport fishers hypothesized that other causal factors besides fishing pressure may have caused the decline. Suspected potential sources of stressors to the CP region included landuse practices, industrial operations, recreational activities, nearshore and shoreline structures, and shipping activities.

ECOLOGICAL RISK ASSESSMENT

1999 Screening Level Ecological Risk Assessment

The 1999 screening-level ecological risk assessment conducted for WDNR specifically focused on identifying and evaluating the natural and anthropogenic factors or stressors that may have contributed to the decline of the CP herring stock (EVS 1999). It also assessed what potential adverse impacts there might be to the stock if additional shipping piers were constructed at Cherry Point.

The screening-level EcoRA concluded that increasing predation and other trends linked to oceanographic conditions had caused increased mortality in the largest and oldest CP herring. After examining the individual stressors and human activities occurring in the CP area, the report concluded that their contribution to the decline of CP herring was negligible. However, potential cumulative effects from these stressors and activities were not assessed due to the lack of available data (EVS 1999). The report also concluded that the impacts of constructing an extension to an existing pier at Cherry Point would have none or only negligible physical and biological effects on the CP herring and their spawning habitat. The only stressors that were identified to have the largest potential impact on the herring were vessel traffic and ballast water that would substantially increase with the construction of the pier.

WDNR utilized the results of this study to reevaluate its leases and other decisions in the CP region. Subsequent decisions and permits that come up for renewal are subject to limitations on aquatic landuses (e.g., limiting vessel traffic to a pier during herring spawning season). WDNR also requires adaptive management in all refinery operations at Cherry Point and requires that fees be used for on-site restoration, mitigation, field studies, and monitoring.

In 2000, WDNR designated the CP region an aquatic reserve as part of its new management plan for sensitive lands and resources. WDNR's Aquatic Reserves Program, in conjunction with its other management efforts, is intended to maintain natural biodiversity, protect and restore ecosystem functions, and maintain appropriate public access to aquatic lands for scientific, educational, and recreational uses. The program is still being developed and is currently under review in accordance with the State Environmental Policy Act (SEPA) process. Once the SEPA process is completed the program will be implemented, and WDNR will develop policies and procedures for its Aquatic Reserves Program.

2000 Regional-Scale Ecological Risk Assessment

WDNR Decision-Making and Management Needs

The results of the 1999 screening-level EcoRA helped WDNR to identify more clearly its decision-making needs and future management goals for effectively managing its ecological resources in the CP region. Those goals were to:

1. Halt the decline of herring in the CP area
2. Reverse the downward trend — bring back the herring
3. Manage the restored herring stock at sustainable levels

WDNR also identified two management needs that would enable it to protect and manage those ecological resources throughout Washington state that are currently under its direct control. Those needs were to:

1. Identify a method to evaluate cumulative impacts to ecological resources
2. Determine whether management options under their control can or cannot restore and sustain natural resources that are in decline such as the CP herring stock

To specifically address these decision-making needs and management goals, WDNR contracted to have a more comprehensive, large-scale EcoRA conducted in 2000 (Landis et al. 2000). The objectives of this regional-scale EcoRA using the RRM developed by Landis and Wiegers (1997) were to:

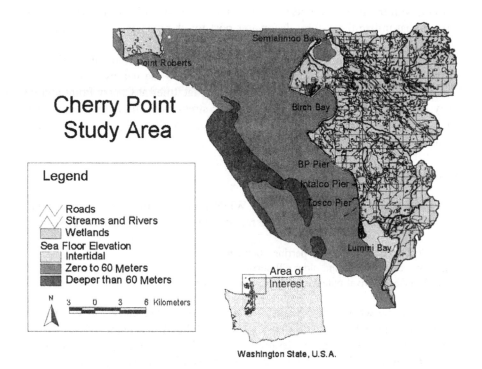

Figure 11.3   The Cherry Point, Washington study area. (From Whatcom County Planning and Development Services, 2001.)

1. Support decision-making needs and goals of Washington DNR in managing its ecological resources in the CP region
2. Conduct retrospective and prospective regional EcoRA of CP herring using RRM
3. Generate risk hypotheses; suggest experimental or field methods to test each hypothesis
4. Identify alternative assessment endpoints in the CP region more appropriate for monitoring and decision making

CP Study Area

The CP study area encompasses the region extending south from the United States–Canada border, including Point Roberts, to the southern tip of Lummi Island and Hale Passage to the south (Figure 11.3). The study area was specifically delineated to include the nearshore areas that were once used as spawning sites by the CP herring stock during its peak abundance in the early to mid-1970s. The entire study area is approximately 715 km^2 and also incorporated the nearshore watersheds that drain into Semiahmoo Bay, Drayton Harbor, Birch Bay, the Strait of Georgia, and Lummi Bay, as well as their respective inter- and subtidal zones.

The nearshore habitats along the coastline are ecologically important for many species of fish, including salmonids and Pacific herring, as well as marine invertebrates, sea and shore birds, and marine mammals (EVS 1999). The region is also economically important with the two oil refineries and an aluminum plant in operation along the CP reach that utilize the natural deepwater port for shipping products to and from their respective piers (EVS 1999).

The upland area is moderately developed with both agricultural and residential landuse occurring in nearby watersheds that drain into the nearby coastal waters (Whatcom County Assessor 2000). The industrial facilities at Cherry Point and the nearby landuses introduce many anthropogenic stressors to the marine ecosystem, including point and nonpoint sources of pollution, as well as physical disturbances such as beach starvation and sedimentation.

Assessment Endpoints

The assessment endpoints selected by WDNR for the initial risk assessment with input from the other stakeholders for the retrospective and prospective EcoRAs were:

1. Continued decline in CP herring abundance since 1973
2. Loss of all 4-year and older CP herring age classes
3. Reduced survival potential of CP herring larvae

To determine what stressors and risk factors could have caused or contributed to the assessment endpoints, an extensive review of the 1999 screening-level EcoRA (EVS 1999) and the scientific literature was conducted. Those stressors and risk factors that were believed to have a high probability of risk or impact to the assessment endpoints were restated as preliminary hypotheses and compiled into a list. In the retrospective EcoRA only historical risk factors and stressors were considered, whereas in the prospective EcoRA current and potential future risk factors and stressors were considered.

Hypotheses

The hypotheses formulated to explain the decline in CP herring stock, loss of older age fish, and reduced larval survival were then organized into categories that addressed large-scale climatic and regional processes, as well as small-scale contaminant effects in relation to the CP herring's life cycle and local population dynamics. The hypotheses were categorized in relation to potential impacts from offshore, nearshore, and on-land sources to the herring. They were not prioritized, nor did the order of listing imply a ranking of probability. The hypotheses were as follows:

Offshore Environments

1. Stochastic population dynamics and ecosystem processes have naturally reduced the CP herring population.
2. Overfishing and overharvesting served as the catalyst to compromise CP herring population survival.
   a. Cessation of harvesting from 1983 through 1986 (4 years) was of insufficient duration for the CP herring stock to fully recover.

      b. Historical harvest levels of 20% and higher for CP herring are too high for the population to support.
3. Lack of recruitment served as the catalyst to compromise CP herring population survival (or a combination of both overfishing and lack of recruitment).
4. Short- and long-term climatic changes are a leading cause of CP herring declines by affecting oceanic conditions, biotic communities, food webs, species distributions, interactions, and survival.
      a. Changes in sea surface temperature affect herring fecundity, survival, and distribution and are a leading cause of declining CP herring.
      b. Changes in salinity affect viability and survival of herring eggs and larvae, causing a decline in CP herring.
      c. Increased predation from salmon, seals, birds, whales, hatchery fish, etc. is a contributing factor to reducing CP herring.
      d. Increased competition with other forage fish, British Columbia (BC) herring, and hatchery fish for food and habitat is a contributing factor to reducing CP herring.
      e. Food supply for larvae, juveniles, and adults has changed in quantity, composition, or quality, leading to CP herring declines.
5. Disease from *Ichthyosporidium hoferi* or other pathogens has compromised the viability of the CP herring and is causing their decline.

Nearshore Environment

6. Habitat loss or degraded habitat cause 4-year-old and older age class herring to decline or avoid the CP area.
      a. On-land and shoreline development have reduced quantity and quality of spawning habitat, making the site less preferable to 4-year-old and older age class herring to use as spawning habitat.
      b. Habitat vegetation or substrate used for spawning has changed in quantity or quality, leading to CP herring declines.
7. Contaminants from vessel traffic, ballast water, and accidental spills are causing the population decline by impacting older age classes or causing avoidance behavior of CP herring from the spawning grounds.
8. Hatchery fish releases (by U.S., Canada, Native American, and Canadian First Nation tribal hatchery operations) and fish plantings (e.g., Atlantic salmon) are directly impacting CP herring by competing for food and habitat, as well as preying on them as food fish.
9. CP herring stock is part of a B.C. herring stock, i.e., the CP herring are a satellite population of a B.C. regional population. The herring are returning to their home range where conditions (food, habitat) are more conducive to 3-year-old and older age class herring.
10. CP herring are a separate population from B.C. herring and are declining due to population and environmentally specific causes. (CP herring are a separate population from the B.C. region population.)

On-Land Environments

11. Contaminants from landuse practices, landfills, contaminated groundwater, permitted NPDES discharges, atmospheric deposition, stormwater runoff, failing septic systems, shoreline development, and recreational activities are causing the CP herring decline by impacting older age classes or causing avoidance behavior of CP herring from nearshore spawning grounds.

Each hypothesis was evaluated using the RRM process whereby numerical ranks are assigned to the specific risk factors and stressors hypothesized based on compiled information and data obtained in the literature review. An in-depth description of the ranks and procedures used in the RRM to identify those factors and stressors with the highest probability of causing the assessment endpoints is presented in the following chapter and in Landis et al. (2004).

RRM Ecological Risk Assessment Results Summary

The retrospective RRM assessment results indicated that offshore stressors, i.e., overexploitation (including overfishing of adults and overharvesting of roe), an associated decrease in recruitment, and changes in oceanic conditions (a warm Pacific decadal oscillation [PDO] regime) during the last 30 years ranked highest as impacting all herring life stages in the CP study area. Nearshore stressors were the second most important source of potential risk to Cherry Point herring, especially to newly hatched larval and juvenile herring. On-land activities, including landuse practices, out-ranked uses related to both nearshore and offshore activities in terms of potential risk from contaminants, especially in the vicinity of the three refineries and their respective piers. There is high probability that the ongoing operations and shipping/cargo-handling activities of the refineries have probably generated ongoing chronic stressors that may have affected the habitat along the Cherry Point reach; however, no direct evidence was found to link contaminant stressors to the decline of the herring stock.

In conducting the prospective analysis it was assumed that no fishing or harvesting of the CP herring would occur for a period of at least 6 to 10 years. It was also assumed that the region would be exposed to the same stressors and uses with the same high probability that there would be an effect associated with the exposure. The results indicate that the highest source of risk from offshore to the CP herring stock will be from climatic changes, specifically another warm PDO event. Nearshore and on-land uses will determine the number and duration of future stressors to the CP herring. Continued industrial development in the CP vicinity, coupled with associated increases in cargo and bulk shipping to the piers, will increase the potential risks to the remaining herring that are currently spawning there. This source of potential risk can be mitigated by best management practices to reduce waste-stream concentrations and by developing long-term, collaborative partnerships among stakeholders in the area with guidance from the appropriate state agencies.

ALTERNATIVE ENDPOINTS

To ensure responsible management of the ecological resources in the Cherry Point region, additional information regarding the chemical, toxicological, organismic, and trophic interactions specific to the Cherry Point area is required. It is understood that severe declines in Puget Sound herring populations have the potential to directly impact the abundance and survival of other valued marine populations in the region.

As a regional-scale assessment endpoint, Pacific herring are both ecologically and economically relevant to the Puget Sound region; however, due to the migratory behavior of Pacific herring, they spend only a few weeks out of the year in their respective spawning areas and are, therefore, less relevant as a site-specific or area assessment endpoint species. The use of other resident species as assessment endpoints specific to the Cherry Point region would be more relevant for monitoring to assess potential risks to the ecological structural and functional integrity of the region. Using that data in combination with Pacific herring data will provide a more comprehensive and realistic evaluation of how these organisms interact, as well as identify the potential risks to those linkages from future impacts. Consequently, a third regional-scale EcoRA was conducted that used alternative endpoints and focused on the nearshore marine environment at Cherry Point (Hart Hayes and Landis 2004). The results of this EcoRA are presented in Chapter 13.

Summary of the Alternative Endpoint Study

The objectives of this study were threefold: (1) to analyze cumulative impacts from multiple sources of chemical and nonchemical stressors in the nearshore region and upland watersheds of Cherry Point, (2) to determine the utility of Monte Carlo type uncertainty analysis in a rank-based regional risk assessment, and (3) to investigate the effects of model habitat characterization on risk estimates.

Several indigenous species in the Cherry Point region were selected as potential alternative assessment endpoints based on their ecological relevance, susceptibility to known or potential stressors, and relevance to WDNR management goals (Table 11.1). These species reflect important characteristics of the Cherry Point environment and are functionally linked to each other, as well as to the Pacific herring both spatially and temporally. The species selected exhibit one or more of the following characteristics:

- Utilize different habitats than the Pacific herring spawning stock
- Utilize habitat with a high probability of exposure to contaminants
- Reside in the region year-round
- Commercially important
- Ecologically connected to the Pacific herring

The information that these alternative endpoints provided in the EcoRA helped identify more specific sources of risk factors and stressors related to landuses, shoreline development, piers and pier construction, shipping, and contaminant inputs occurring in the CP area. The results of this regional-scale EcoRA are described in greater detail in Chapter 13 of this volume; however, they did indicate that the major contributors of risk to the biota in the CP region are vessel traffic, upland,.urban and agricultural landuse, and shoreline recreational activities. The results of this study in combination with the retrospective and prospective EcoRAs conducted for the CP study area using the RRM will help facilitate WDNR's ability to make decisions and thereby help WDNR to achieve its management goals for the CP region.

Table 11.1　Alternative Assessment Endpoints and Criteria Used for Selection

Assessment Endpoints	Reasons for Selection
Eelgrass (*Zostera marina*) and marine macroalgae	1　Keystone species for detritus-based food webs in nearshore habitats. Structural habitat increases substrate surface area for epiphytic algae and associated microzooplankton, reduces wave and current action, traps sediment and other detritus, maintains high dissolved oxygen concentrations, and mitigates temperature fluctuations by providing shade (Kikuchi and Peres 1977).
	2　Provides habitat to a variety of nearshore species including Pacific herring and Dungeness crab.
	3　Juvenile herring feed almost exclusively on the invertebrates that inhabit the eelgrass beds (Levings 1983).
	4　Eelgrass is the preferred spawning substrate for Pacific herring.
Coho salmon (*Oncorhyncus kisutch*)	1　Utilizes estuarine and nearshore habitat as juveniles.
	2　Utilizes stream habitat for spawning.
	3　Juveniles compete with Pacific herring for prey and older juveniles and adults prey upon Pacific herring.
Juvenile English sole (*Pleuronectes vetulus*)	1　Juveniles utilize nearshore region of study area.
	2　Bottom dwelling: high probability of exposure to and, therefore, effects from toxicants present in the sediments.
	3　Known reproductive and cancerous effects from toxicants in sediments.
Surf smelt (*Hypomesus pretiosus*)	1　Like Pacific herring, surf smelt are a forage fish.
	2　Utilizes sand and gravel beaches for spawning habitat — high probability of exposure to and, therefore, effects from toxicants present in the sediments.
Juvenile Dungeness crab (*Cancer magister*)	1　Commercially important.
	2　Utilize sediment as habitat — high probability of exposure to and, therefore, effects from toxicants present in the sediments.
	3　Utilizes eelgrass and marine macroalgae for cover as juveniles.
Littleneck clams (*Protothaca staminea*)	1　Commercially important.
	2　Preyed upon by Dungeness crab.
	3　Utilize sediment as habitat — high probability of exposure to and, therefore, effects from toxicants present in the sediments.
Great blue heron (*Ardea herodius*)	1　Link between terrestrial and aquatic habitat.
	2　Preys upon Pacific herring, surf smelt, English sole, and crustaceans as well as terrestrial rodents.
	3　Forages in eelgrass beds.

REFERENCES

Bargmann, G. 1998. State of Washington Forage Fish Management Plan. Washington Department of Fish and Wildlife, Fish Management Program, Marine Resources Division, Olympia, WA, 65 pp.

EVS Environmental Consultants, LTD. 1999. Cherry Point Screening Level Ecological Risk Assessment. Prepared for the Washington Department of Natural Resources. EVS Project No. 2/868-01.1. EVS Environmental Consultants, Seattle, WA, 465 pp.

Gonyea, G., Burton, S., and Penttila, D. 1982. Summary of 1981 Herring Recruitment Studies in Puget Sound. Washington Department of Fisheries Report No. 157, Olympia, WA, 27 pp.

Hart Hayes, E. and Landis, W.G. 2004. Regional ecological risk assessment of a nearshore marine environment: Cherry Point, WA. *Hum. Ecol. Risk Assess.*, 10, 299–325.

Kikuchi, T. and Peres, J.M. 1977. Consumer ecology of seagrass beds, in *Seagrass Ecosystems: A Scientific Perspective*, McRoy, C.P. and Herfferich, C., Eds., Marcel Dekker, New York, 147–193.

Kyte, M.A. 2000. Observations on a Qualitative Examination of the Intertidal Zone of the Cherry Point Shoreline, 3 and 4 June 2000. Prepared for BP Cherry Point Refinery, Alcoa Intalco Works and Tosco Ferndale Refinery. Report No. 003-1327-000.

Kyte, M.A. 2001. Observations from a Qualitative Examination of the Intertidal Zone of the Cherry Point Reach June 22–24, 2001. Prepared for BP Cherry Point Refinery, Alcoa Intalco Works and Tosco Ferndale Refinery. Report No. 003-1327-000.

Landis, W.G. and Wiegers, J.A. 1997. Design considerations and a suggested approach for regional and comparative ecological risk assessment, *Hum. Ecol. Risk Assess.*, 3, 287–297.

Landis, W.G., Duncan, P.B., Hart Hayes, E., Markiewicz, A.J., and Thomas, J.F. 2004. A regional assessment of the potential stressors causing the decline of the Cherry Point Pacific herring run and alternative management endpoints for the Cherry Point Reserve (Washington, USA), *Hum. Ecol. Risk Assess.*, 10, 271–279.

Landis, W.G., Markiewicz, A.J., Thomas, J., and Bruce, D.P. 2000. Regional Risk Assessment for the Cherry Point Herring Stock. Report prepared for the Washington Department of Natural Resources. Prepared by the Institute of Environmental Toxicology, Western Washington University, Bellingham, 83 pp.

Lemberg, N.A., O'Toole, M.F., Penttila, D.E., and Stick, K.C. 1997. Washington Department of Fish and Wildlife 1996 Forage Fish Stock Status Report. Washington Department of Fish & Wild. Stock Status Report No. 98, Olympia, WA, 83 pp.

Levings, C.D. 1983. Some observations of juvenile herring at the Fraser River estuary, B.C., Proceedings of the Fourth Pacific Coast Herring Workshop, October 1981, in *Can. Manuscr. Rep. Fish. Aquat. Sci.*, 1700, 91–103.

Penttila, D. and Burton, S. 1986a. Summary of 1984 Herring Recruitment Studies in Puget Sound. State of Washington Department of Fisheries Progress Report No. 256, November 1986. Washington Department of Fish and Wildlife, Olympia, WA, 42 pp.

Penttila, D., Burton, S., and M. O'Toole, M. 1986b. Summary of 1985 Herring Recruitment Studies in Puget Sound. State of Washington Department of Fisheries Progress Report No. 257, October 1986. Washington Department of Fish and Wildlife, Olympia, WA, 42 pp.

Penttila, D., Burton, S., and Gonyea, G. 1985. Summary of 1983 Herring Recruitment Studies in Puget Sound. State of Washington Department of Fisheries Progress Report No. 223. Washington Department of Fish and Wildlife, Olympia, WA, 42 pp.

Stick, K. 1990. Summary of the 1989 Pacific herring spawning ground surveys in Washington State Waters. Washington State Department of Fish and Wildlife Progress Report No. 280, July 1990.

WDNR (Washington Department of Natural Resources). 2001. DNR Aquatic Mission. http://www.wa.gov/dnr/htdocs/aqr/html (accessed October 25, 2001).

Whatcom County Assessor. 2000. 2000 tax parcels [computer file]. Whatcom County Assessor, Bellingham, WA.

Whatcom County Planning and Development Services. 2001. Water Resource Inventory Area 1 Watershed boundaries [computer file]. Whatcom County Planning and Development Services, Bellingham, WA.

CHAPTER 12

Retrospective Regional Risk Assessment Predictions and the Application of a Monte Carlo Analysis for the Decline of the Cherry Point Herring Stock

Wayne G. Landis, Emily Hart Hayes, and April M. Markiewicz

CONTENTS

INTRODUCTION

The history of the Pacific herring stock at Cherry Point, Washington and a series of alternative hypotheses for its decline were presented in Chapter 11. We (Landis et al. 2004) conducted a regional ecological risk assessment using the relative risk model (RRM) to investigate the causes of the current decline, current risks to the

1-56670-655-6/04/$0.00+$1.50
© 2004 by CRC Press LLC

population, and the outcomes of future management options. The population decline of the herring corresponds to a collapse of the age structure, although survivorship of eggs to the age 2 class has not diminished. The range of spawning areas has also declined, with the area of Point Whitehorn as the principal location.

The retrospective risk assessment identified climate change, as expressed by the warmer sea surface temperatures associated with a warm Pacific decadal oscillation (PDO) and exploitation as important risk factors. The warmer water also changes patterns in food resources, predators, and water quality. Contaminants have the potential for impact, but exposure to the eggs, hatchlings, and fry has not been demonstrated at Cherry Point (CP). Exposure of adults to contaminants during migration may occur and has been included into our assessment. Modeling of the population age vs. fecundity curves and survivorship data indicate that the current population of ages 2 and 3 fish cannot be self-sustaining without the survivorship or immigration of age 4 and older fish.

Because of the limitations on the available data for a large number of the stressors and the stressor–habitat–impact relationship, there is a great deal of uncertainty associated with this assessment. Data at a comparable regional scale to that for the PDO are not available for contaminants, fishing pressure, disease, and other potential causative agents. This leads to a great deal of uncertainty. As we began to analyze this uncertainty by applying Monte Carlo techniques it readily became apparent that the retrospective RRM–Monte Carlo synthesis is essentially a quantitative weight-of-evidence (WoE) approach.

The approach described below combines WoE and causality criteria with a multitude of stressors at a regional scale. The difficulties include how to deal with differences in the magnitude of effects and how to express the uncertainty as distributions.

We applied a WoE and path analysis approach based upon our RRM in order to estimate the cause of the decline of the CP Pacific herring. This WoE approach is based upon a risk assessment type conceptual model in order to link the paths of potential sources of stressors to the effects seen in the population. Ranking criteria and regressions are used to assign weights to the potential sources and stressors. A Monte Carlo analysis is applied to represent the uncertainty in each of the ranks, correlations, and filters and to estimate the uncertainty of the analysis. This technique results in a series of multinomial distributions representing the likelihood of a stressor causing an impact. In the case of the CP herring, climate change, habitat alteration, and contamination at a landscape scale were identified as important stressors. This case study demonstrates that a clearly derived and quantified WoE and path analysis approach is useful to investigating casual links at regional scales.

RETROSPECTIVE RRM AND WOE SYNTHESIS

The difficulty with the retrospective analysis is that it is very difficult to quantify the uncertainty with this type of procedure. In order to better describe the uncertainty with the assignment of probable cause it is important to investigate other methods. The WoE approach as outlined by Menzie et al. (1996) is a promising approach.

Weight of Evidence

Classic methodologies such as Hume's criteria and Koch's postulates do not work well for open systems with diverse symptoms. The open system and the large scale associated with sites such as Cherry Point preclude experimentation. Large-scale factors such as the PDO are not possible to manipulate and must be incorporated into any causal framework. Ecoepidemiological approaches such as those of Suter et al. (2002) are not inherently quantitative, but rely or scoring schemes that are not easily manipulated mathematically and that do not incorporate uncertainty. The quantification of the scoring scheme and the express statement of uncertainty are both important factors in a useful means of assigning causality.

A retrospective assessment coupled with a modern idea of scale, WoE approach and uncertainty analysis can produce a quantitative framework for ranking risk factors. The next paragraphs describe how the RRM was modified for a retrospective assessment incorporating Monte Carlo analysis to describe uncertainty.

RELATIVE RISK MODEL AND THE WOE APPROACH

The RRM was developed during our ecological risk assessment of Port Valdez, Alaska. Like this study area, Port Valdez has a variety of anthropogenic stressors including fish hatcheries, fish processing wastes, petroleum-based effluents from the pipelines, municipal effluents, and tanker traffic (Landis and Wiegers 1997; Wiegers et al. 1998). The variety of stressors and endpoints led Wiegers and colleagues to the source–habitat–impact model for conceptual model development. This approach is as described in Chapter 2 and is briefly summarized in the following paragraphs.

SOURCE–HABITAT–IMPACT

In a regional multiple-stressor assessment, the number of possible interactions increases exponentially. Stressors arise from diverse sources, receptors are associated with a variety of habitats, and one impact may lead to additional direct and indirect effects. The approach of our current regional assessment model is to identify the sources and habitats in different locations (risk regions) of the Cherry Point coastal system, rank their importance in each location, and combine this information to predict relative levels of risk. The number of possible risk combinations resulting from this approach depends on the number of groups identified in each risk region. For example, if two source types and two habitat types are identified, then four possible combinations of these components can lead to an impact. If we are concerned about two different impacts, eight possible combinations exist.

Use of Ranks and Filters to Quantify Relative Risk

Our regional approach incorporates a system of numerical ranks and weighting factors to address the difficulties encountered when attempting to combine different kinds of risks. Ranks and weighting factors are unitless measures that operate under different limitations than measurements with units (e.g., mg/L, individuals/cm^2). We link these ranks to specific locations within a landscape, providing a map of risks with the sources of risk clearly identified.

Spatially Explicit

Sources and habitats are specifically included in the risk assessment, making it spatially explicit. Risks can be defined for specific areas, within the context of the entire region. Gradients of risk may exist due to the presence of a variety of stressors generated by a variety of sources. The relative risks can be mapped and decisions made at a regional level.

Use in a Prospective and Retrospective Approach

Previously published studies include examples of prospective risk assessments where future impacts are calculated. In a retrospective risk assessment the goal is to identify stressors and the sources that have contributed to an observed historical impact in that environment. The process reverses the normal order of consideration from source–habitat–impact to impact–habitat–source.

Common Ranking Methodology

The numerical scores that are obtained in the ranking process are unique to the set of decisions and ranking criteria derived for that specific region. The numerical scores cannot be compared directly to other studies or regions unless a set of newly derived scoring procedures is derived. If several areas are being compared in order to set remediation or management priorities, then each area needs to be combined into a single RRM setting. This approach provides the setting for the analysis of the cause of the decline of the Pacific herring stock at Cherry Point.

RETROSPECTIVE WOE ANALYSIS FOR CHERRY POINT PACIFIC HERRING

The basic conceptual model for the CP Pacific herring has been adapted for this retrospective analysis (Figure 12.1). The conceptual model incorporates each of the sources providing the stressors linked to the observed impacts in the population. A simplified model that deals only with the source climate change is presented in Figure 12.2.

The source of the change in temperature within the northeastern Pacific Ocean is known as the PDO. Climate change is also a source of habitat alteration as species migrate because of alterations in conditions. Predators may increase or decrease in number, nutrient fluxes can be altered, and the distribution of prey items changed. Disease may also be an important issue as new pathogens may be brought in by the

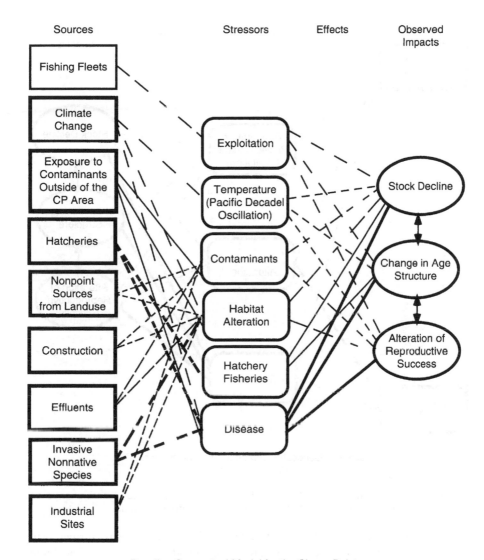

Baseline Conceptual Model for the Cherry Point
Pacific Herring

Figure 12.1   Conceptual model for the WoE approach to determining the likely cause of the stock decline at Cherry Point.

change in conditions. In the simplified model there are a total of 13 interactions that must be considered.

The next stage is that the evidence for each source, stressor, and potential linkage is examined and provided a rank with a description of the associated uncertainty. Table 12.1 provides examples of such a process with three example sources. In the current example, all the ranks and descriptions are based upon Landis et al. (2004). In the next step the available information is used to assign a distribution to each rank depending upon the uncertainty associated with each factor.

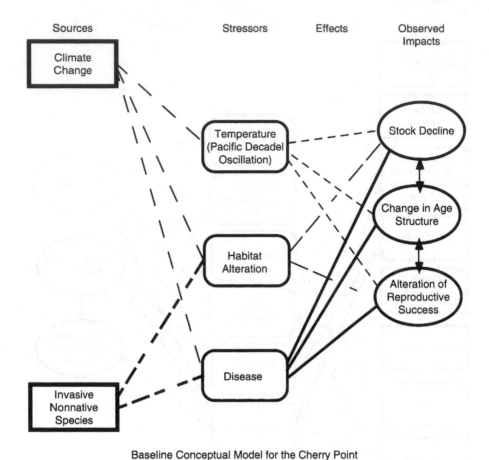

Baseline Conceptual Model for the Cherry Point
Pacific Herring

Figure 12.2 Simplified conceptual model identifying the links due to climate change and invasive nonnative species as the source of the stressors.

After the assigning of distributions, a Monte Carlo analysis is performed using Crystal Ball® 2000 software as a macro in Microsoft® Excel 2002. The Monte Carlo simulations were run for 1000 iterations, and output distributions for each subregion, source, habitat, and endpoint risk prediction were derived. The distributions depict a range of probable risk estimates associated with each point estimate. After running preliminary simulations of up to 10,000 iterations, 1000 iterations appeared sufficient and resulted in similar results. This procedure allows us to estimate the resultant uncertainty in the retrospective analysis.

Figure 12.3 illustrates the distribution selected to represent the uncertainty for three variables. In the case of fishing fleets as a source, there is a great deal of documentation that fishing of Pacific herring, both at large scales and upon the spawning fish, has occurred. The uncertainty is in the fact that it is not clear what amount of offshore fishing has directly affected the Cherry Point Pacific herring and how much has been on stocks that are not related. In this case a rank of 6 is the

Table 12.1   Examples of Source Ranks, Notes, and Uncertainty Description

Source	Rank	Notes on Initial Ranking	Uncertainty Description	Uncertainty
Fishing fleets and nearshore fisheries	6	Fishing fleets are well documented within U.S. waters until the 1980s and currently exist in Canadian waters.	It is clear that CP Pacific herring have been fished in the past, but no fishing is currently allowed in Washington waters. Take in Canadian waters is possible, and Pacific herring from Puget Sound have been recovered in Canadian waters.	Low
Nonpoint sources from landuse	2	Nonpoint sources do occur but much of the land is forested or otherwise covered. Areas in the Point Whitehorn region have become residential.	Nonpoint sources are more difficult to map and uncertainty exists in the landuse classifications for the CP region.	Moderate
Exposure to contaminants outside of the CP area	4	Contaminants outside of the Cherry Point region are known to exist from numerous examinations of heron eggs, marine mammal blubber, and fish. Many of the compounds are legacy pollutants such as derivatives of DDT, PCBs, and members of the dioxin and furan classes	High uncertainty because CP Pacific herring and other fish species in the area have not been sampled before coming to the CP area. Many potentially toxic materials have not been analyzed, especially the halogenated organics or other estrogen disruptors.	High

preferred input, with a probability of 0.80, but a rank of 4 is also given a set probability (0.20).

An intermediate case is effluents. Effluents are common throughout the Georgia Straits and Puget Sound region, but toxicity is not generally high and rapid dilution occurs because of the magnitude of the currents in the region. However, local high concentrations from industrial sources or untreated stormwater runoff could be damaging. In this instance the most common rank is a 4, with ranks of 2 and 6 given a lower probability.

Invasive species provide a case with a low rank. Although invasive species do exist in the broad geographic region, they are not particularly prevalent in the habitat used by Pacific herring. However, lack of evidence may also be because there has not been an extensive survey with the region for these types of organisms. So, the initial ranking and the rank given the source are a 2, but a rank of 0 and a rank of 4 are given equal probabilities.

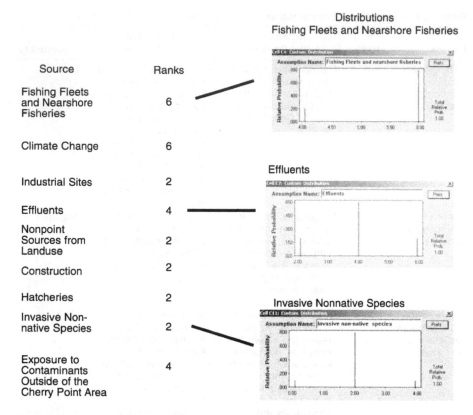

Figure 12.3   Example of how distributions are established for the various ranks in the conceptual model

The linkages (filters) between the source–habitat–impact pathways are also assigned a probability. A value of 1.0 means that such exposure or a causal pathway exists. A value of 0.0 means that no exposure or mechanism of impact exists for that group of relationships. An intermediate case 0.5 can be assigned with a probability distribution if it is not clear if a particular pair of interactions is linked.

When possible the causal criteria as described by Adams (2003) were used to evaluate the values of the filters (Table 12.2). Unfortunately, it is difficult to meet these criteria over the scales relevant to the biology and life history of the Pacific herring in an environment such as the Straits of Georgia with large-scale spatial and temporal relationships. Because it is not clear what the migration paths of the Cherry Point herring are, nor the genetic relationships to other stocks, or even how the current stock is representative of that of the early 1970s, there is a lot of room for uncertainty.

After the ranks are assigned for each source and stressor and the linkages are assigned an uncertainty, then the Monte Carlo computation is performed. The output is a distribution as portrayed in Figure 12.4.

In these figures the calculated value for the retrospective assessment is marked as a solid line. In Figure 12.4a it can be seen that the distribution is generally below

Table 12.2 Criteria for Causality

Causal Criteria	Description
1. Strength of association	Cause and effect coincide.
2. Consistency of association	The association between a particular stressor or stressors and an effect has been observed by other investigators in similar studies and at other times and places.
3. Specificity of association	The effect is diagnostic of exposure.
4. Time order or temporality	The cause precedes the effect in time, and also the effect decreases when the cause is decreased or removed.
5. Biological gradient	There is a dose–response relationship either spatially or temporally within the system. The risk of an effect is a function of magnitude of exposure.
6. Experimental evidence	Valid experimental studies support the proposed cause–effect relationship.
7. Biological plausibility	There is credible or reasonable biological or toxicological basis for the proposed mechanism linking the proposed cause and effect.

Source: Data from Adams, S.M., *Hum. Ecol. Risk Assess.,* 9, 17–35, 2003.

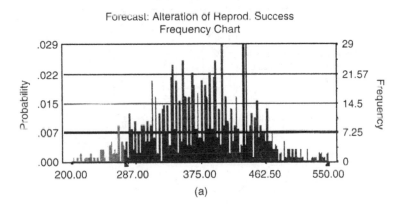

(a)

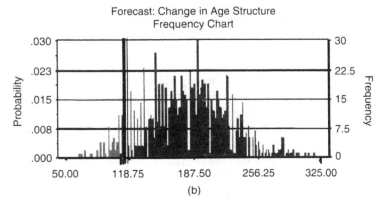

(b)

Figure 12.4 Distributions for the forecast of two of the observed effects, alteration of reproductive success, and change in age structure

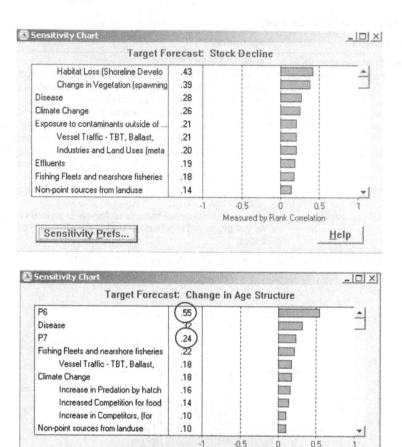

Figure 12.5   The sensitivity analysis points to the importance of better understanding the causal relationships connecting contamination and climate change to a change in the occurrence of disease and a change in the age structure of the population.

the original estimate, indicating that the original may have been an overestimate of the true risk. Figure 12.4b compares the original estimate for risk to the change in age structure to the distribution, and the original seems to be an underestimate of the degree of risk involved. Note that the risk scores for decline of the population are higher than those of age structure. This is because a change in age structure is one of the factors incorporated into the overall population decline.

One of the advantages of the WoE approach using Monte Carlo is that the process allows the examination of what factors within the model drive the final distribution of results. An improvement in the uncertainty associated with these factors should reduce the overall uncertainty of the estimates.

Two examples are found in Figure 12.5. In the sensitivity chart for stock decline, habitat loss and change in vegetation lead the sensitivity scores. Both factors have high ranks, but also have a great deal of associated uncertainty. Note that climate change received a relatively low sensitivity score although it has a high rank. This is because there is little doubt that the PDO occurs and can have important effects so that the input distribution was very narrow. Input factors with low uncertainty are essentially constants and are not a cause of variability in the output.

In the chart ranking sensitivity for the change in age structure, two of the linkages between stressors and effects, P7 and P6, have a high sensitivity. This sensitivity is because many of the mechanisms tying the impacts of contaminants to alterations in age structure are unclear. These linkages were given equal probability of tying a stressor to an effect and have a resultant high sensitivity score.

CONCLUSIONS AND RECOMMENDATIONS

The WoE approach coupled with a Monte Carlo analysis of the uncertainty proved useful and led to a series of conclusions:

- The declines observed at Cherry Point and the age structure common to Puget Sound stocks are due to large-scale events, such as habitat loss and the PDO.
- Contaminants are possibly an important stressor, but there is considerable uncertainty in the linkage of toxicity to changes in age structure and population decline at large scales.
- A WoE approach can incorporate a variety of stressors and pathways and is based upon the diagnostic symptoms observed in the spawning stocks.
- It is possible to incorporate uncertainty in the rankings and linkages into the description of the WoE estimate.
- The WoE combined with the uncertainty analysis approach is flexible, and new stressors and linkages can be easily added or subtracted as evidence is accumulated.

ACKNOWLEDGMENTS

The research was funded by the Washington State Department of Natural Resources.

REFERENCES

Adams, S.M. 2003. Establishing causality between environmental stressors and effects on aquatic systems, *Hum. Ecol. Risk Assess.*, 9, 17–35.

Landis, W.G. and McLaughlin, J.F. 2000. Design criteria and derivation of indicators for ecological position, direction and risk, *Environ. Toxicol. Chem.*, 19, 1059–1065.

Landis, W.G. and Wiegers, J.A. 1997. Design considerations and a suggested approach for regional and comparative ecological risk assessment, *Hum. Ecol. Risk Assess*, 3, 287–297.

Landis, W.G., Duncan, P.B., Hart Hayes, E., Markiewicz, A.J., and Thomas, J.F. 2004. A regional assessment of the potential stressors causing the decline of the Cherry Point Pacific herring run and alternative management endpoints for the Cherry Point Reserve (Washington, USA). *Hum. Ecol. Risk Assess.*, 10, 271–297.

Landis, W.G., Markiewicz, A.J., Thomas, J.F., and Hart Hayes, E. 2002. Regional risk assessment predictions for the decline and future management of the Cherry Point herring stock and region, Proceedings of the 2001 Puget Sound Research Conference. Droscher, T., Ed., Puget Sound Water Quality Action Team, Olympia, Washington. Available at: http://www.psat.wa.gov/Publications/01_preceedings/session/sess5b.htm.

Menzie, C., Henning, M.H., Cura, J., Finkelstein, K., Gentile, J., Maughn, J., Mitchell, D., Petron, S., Potocki, B., Svirsky, S., and Tyler, P. 1996. A weight-of-evidence approach for evaluating ecological risks: report of the Massachusetts Weight-of-Evidence Work Group, *Hum. Ecol. Risk Assess.*, 2(2), 277–304.

Suter, G., Jr., Norton, S., and Cormier, S. 2002. A methodology for inferring the causes of observed impairments in aquatic ecosystems, *Environ. Toxicol. Chem.*, 21, 1101–1111.

Wiegers, J.K., Feder, H.M., Mortensen, L.S., Shaw, D.G., Wilson, V.J., and Landis, W.G. 1998. A regional multiple stressor rank-based ecological risk assessment for the fjord of Port Valdez, AK, *Hum. Ecol. Risk Assess.*, 4, 1125–1173.

CHAPTER 13

Ecological Risk Assessment Using the Relative Risk Model and Incorporating a Monte Carlo Uncertainty Analysis

Emily Hart Hayes and Wayne G. Landis

CONTENTS

We conducted a regional ecological risk assessment for the Cherry Point (CP) region in northern Whatcom County, Washington using the relative risk model (RRM). The study had three objectives: (1) to analyze cumulative impacts from multiple sources of stress to assess risk to multiple biological endpoints that utilize the region, (2) to determine the applicability of the RRM in the study area, and (3) to use Monte Carlo analysis of the uncertainties in the RRM approach.

We used geographic information systems (GIS) to compile and compare spatial data for sources of stressors and habitats in subregions within the study area. These data determined the ranks for each subregion. By quantitatively integrating ranks with exposure and effects filters as defined in a conceptual model, we estimated relative risk in subregions, relative contribution of risk from sources, risk in habitat types, and assessment endpoints most at risk within the CP area. Finally, we used Monte Carlo techniques to perform uncertainty analysis and applied an alternative ranking scheme to evaluate the effects of model and parameter uncertainty on the risk predictions.

The RRM and uncertainty analysis results suggest that the major contributors of risk in the region are commercial and recreational vessel traffic, upland urban and agricultural landuse, and shoreline recreational activities. The biological endpoints most likely to be at risk are great blue heron and juvenile Dungeness crab. The majority of risk occurs in sandy intertidal, eelgrass, and macroalgae habitats. The subregions where the most risk occurs are Lummi Bay, Drayton Harbor, and Cherry Point.

INTRODUCTION

Recent trends in ecological risk assessment have shifted toward assessing risk from multiple stressors at a regional scale (Cook et al. 1999; Cormier et al. 2000). Such regional-scale risk assessments present many benefits and challenges. Regional risk assessments benefit natural resource managers by providing an integrated picture of risk from multiple chemical and nonchemical stressors to aid in decisions that benefit entire regions and ecosystems. The necessity to analyze risk at a regional scale demands risk assessment methods that can account for the many spatial scales at which stressors and endpoints can occur throughout the landscape. The use of chemical- or receptor-specific methods falls short of addressing these multiple spatial scales. The RRM (Landis and Wiegers 1997; Wiegers et al. 1998) provides an alternative to chemical- and receptor-specific methods.

The RRM integrates spatial information into the risk assessment process. Using GIS to analyze spatially explicit datasets, the RRM ranks sources of stressors and habitats for subregions within the study area (Landis and Wiegers 1997; Wiegers et al. 1998). By quantitatively determining the interactions between sources and habitats, the relative risk in subregions, contribution of risk from sources, risk in habitats, and risk to assessment endpoints can be calculated in a region. The ultimate difference between an RRM risk assessment and traditional risk assessments is the depiction of risk in a spatial context, allowing natural resource managers to make decisions based on information about geographically distinct risk.

This chapter describes an application of the RRM for a regional-scale ecological risk assessment of the CP region in northwestern Washington. The study had three objectives: (1) to analyze cumulative impacts from multiple sources of stress to assess risk to multiple biological endpoints that utilize the region, (2) to determine the utility of the RRM applicability in the CP study area, and (3) to use Monte Carlo analysis of the uncertainties in the RRM approach.

PROBLEM FORMULATION

The problem formulation phase began the process of analyzing the effects of multiple stressors on biological endpoints in the CP region. During the problem formulation phase of this assessment, we defined the spatial extent of the study area and subregions, identified sources, stressors, and assessment endpoints, and developed a conceptual model to derive preliminary hypotheses about potential exposure and effects pathways and resulting risk in the CP environment.

Risk Assessment and the Cherry Point Region

The study area consists of the coastline from Point Roberts and the U.S. border in the north to the southern boundary of Lummi Bay in the south (Figure 13.1). The area incorporates approximately 715 km² and includes the nearshore watersheds that drain into Semiahmoo Bay, Birch Bay, Lummi Bay, and the Strait of Georgia as well as the inter- and subtidal regions in these water bodies.

Cherry Point nearshore habitats are ecologically important for many species including several fish, marine invertebrates, sea and shore birds, and marine mammals (EVS 1999). The region is also economically important. Two oil refineries and an aluminum plant maintain shipping piers on the coast (EVS 1999). The upland area is moderately developed with both agricultural and residential landuse occurring in watersheds that drain into coastal waters (Whatcom County Assessor 2000; Whatcom County PUD 2000). The industrial facilities and upland landuses introduce many anthropogenic stressors to the biotic components in the ecosystem, including point and nonpoint sources of pollution, beach sedimentation and sediment starvation, and other physical disturbances.

The Washington Department of Natural Resources (WDNR) manages the aquatic lands of Washington state "for current and future citizens of the state to sustain long-term ecosystem and economic viability" (WDNR 2001a). This joint mission to both protect natural resources and generate income from them creates a framework in which difficult management decisions must be made. The purpose of this regional risk assessment was to provide estimates of the relative contributions of risk from

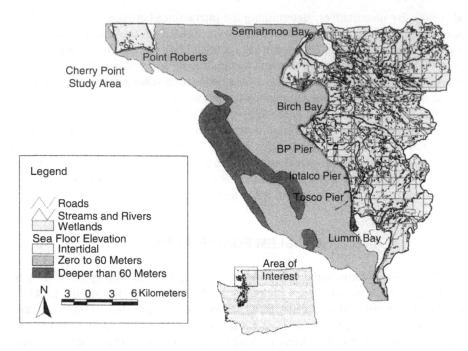

Washington State, U.S.A.

Figure 13.1 The CP study area in Northern Whatcom County, Washington. BP (British Petroleum) Oil Company, Alcoa Intalco Works Aluminum, and Tosco Oil Company maintain shipping piers on the coast. (See color insert following page 178.)

anthropogenic sources to biological endpoints in the CP region to aid WDNR in its management decisions.

Two ecological risk assessments have been conducted in the CP region to determine the effects of chemical and physical stressors on the Pacific herring (*Clupea pallasi*) that spawn each spring at Cherry Point (EVS 1999; Landis et al. 2000a; Markiewicz et al. 2001). This Pacific herring stock has experienced dramatic declines in population size and a compression of age structure since the 1970s (EVS 1999). The assessments concluded that the major risk factors affecting Pacific herring in the CP region are the effects associated with the Pacific decadal oscillation (PDO), a 30-year sea temperature warming and cooling cycle in the Pacific Ocean, and historical overharvesting of fish and roe.

The Pacific herring ecological risk assessments provided valuable information to decision makers about future risks to Pacific herring at Cherry Point; however, they did not provide information about any other species in the region and do not, therefore, provide a complete characterization of the potential risks to the region as a whole. The study area boundaries and risk components selected during the problem formulation phase of this assessment were specifically chosen in an attempt to provide a multiple endpoint risk characterization as an alternative to the herring-specific risk assessments for the CP region.

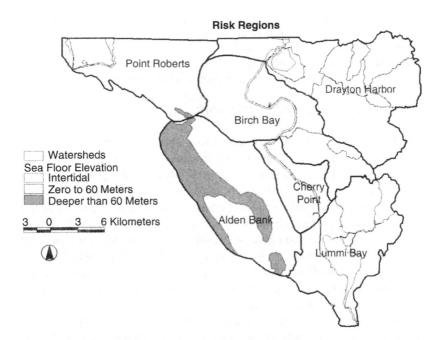

Figure 13.2 The study area divided into six subregions based on watershed and bathymetric boundaries. (See color insert following page 178.)

Study Area and Subregions

Using ArcView™ GIS software, we defined the boundaries of the study area and divided it into six subregions (Figure 13.2) based on watershed and bathymetric boundaries (Whatcom County Planning and Development Services 2001; NOS 2001) and the location of the recently established WNDR aquatic reserve (WDNR 2001b). Upland, the study area ends at the boundaries of watersheds draining directly into coastal waters. Nearshore, the study area was limited to waters within the 60-m contour, representing the depth of waters where assessment endpoint species are most likely to be found (Laroche and Holton 1979; Krygier and Pearcy 1986; Gunderson et al. 1990; Shi et al. 1997). The six subregions are (Figure 13.2):

1. Point Roberts subregion, consisting of Point Roberts proper, a peninsula protruding into the northern boundary of the study area immediately south of the U.S.–Canadian border, plus the adjacent waters to 60-m depth
2. Drayton Harbor subregion, comprising Drayton Harbor itself and the watersheds that drain into this water body including the city of Blaine, California and Dakota Creeks, Semiahmoo Spit, and adjacent waters
3. Birch Bay subregion, containing the bay and Birch Bay State Park, Terrell Lake, Terrell Creek, and the remaining upland watershed
4. CP subregion, which includes the newly designated CP aquatic reserve, three large industrial piers and much of the upland industrial complexes, the site of a proposed pier and shipping facility, as well as several small unnamed creeks

5. Lummi Bay subregion, consisting of Lummi River, part of the city of Ferndale, a large portion of the southern oil refinery complex, the Lummi Nation Indian Reservation, and Lummi Bay itself
6. Alden Bank subregion, an offshore area with no terrestrial component and centered around a shallow bank that rises from deeper waters closer to shore

Identification of Assessment Endpoints

The Cherry Point Technical Working Group, organized by the WDNR Aquatic Resources Division, represented stakeholders for the endpoint selection process. The working group included representatives from WDNR, Washington Department of Fish and Wildlife (WDFW), Washington State Department of Ecology (Ecology), the Lummi Nation Indian tribe, citizens' groups, and the three major industries in the region (British Petroleum oil refinery, Alcoa Intalco Works, and Phillips 66 oil refinery). This stakeholder group generated a list of species based on accepted criteria for the selection of assessment endpoints (Suter 1993; USEPA 1998). The list was then shortened to six biological endpoints that included representative components of the CP ecosystem, paying special attention to select endpoints that are susceptible to site-specific stressors in the CP region.

Another important factor in refining the stakeholder list of endpoints to those appropriate for use in the study included a careful examination of spatial scales of species vs. the spatial extent of the study area. The size and boundaries of the study area were designated according to the spatial scale of WDNR management decisions. However, because some of the life stages of potential assessment endpoints extend far beyond the boundaries of the study area, care was taken to limit the study to life stages in which spatial scales match the spatial extent of the study area. The selected assessment endpoints include three fish (Coho salmon, juvenile English sole, and surf smelt embryos), two macroinvertebrates (juvenile Dungeness crab and adult native littleneck clam), and one bird (great blue heron). Care should be taken, therefore, to avoid extending endpoint risk predictions as representative of all life stages if only a given life stage is specified as the assessment endpoint. Risk to the juvenile life stage of Dungeness crab is not equivalent to risk to Dungeness crab larvae or Dungeness crab adults and should not be misconstrued as such.

Coho salmon are known to utilize nearshore and stream habitats in the study area (Miller et al. 1977; NSEA 2000) and are culturally valued by stakeholders. Coho salmon are connected to Pacific herring in the marine food web via predator–prey relationships and competition for food. Juvenile coho compete with Pacific herring for prey, and older juveniles and adults prey upon Pacific herring (Healey et al. 1980; Holtby et al. 1990; Brodeur and Pearcy 1992). Juvenile English sole are also known to use the nearshore region at Cherry Point (Kyte 1993; Kyte 1994). Because the juvenile life stage is benthic, sole are likely to be exposed to and exhibit effects from contaminated sediments (Malins et al. 1985; Rhodes and Casillas 1985; Johnson et al. 1988; Stein et al. 1991; Collier et al. 1992; Johnson et al. 1993; Johnson et al. 1998; Myers et al. 1998; Johnson et al. 1999). If contaminated sediments are present in the study region in a high enough concentration, English sole would likely exhibit a response.

Pacific surf smelt embryos have been documented, and the species is known to spawn year-round on beaches within the CP study area (Pentilla 1997; WDFW 2002a). The close association of surf smelt embryos with sediments makes them vulnerable to potential stressors in the region, such as contaminated sediments, anoxia, and changes in sediment composition (Chapman et al. 1985; Hirose and Kawaguchi 1998). Surf smelt also support both commercial and recreational fisheries in the state (WDFW 2002a), making them important to local stakeholders.

The juvenile life stage Dungeness crab are known to inhabit nearshore waters in the CP study area, as well (McMillan 1991). Like English sole, their close association with sediments makes them vulnerable to potential stressors in the region, including sediment changes and contaminants. Contaminated water and sediments affect Dungeness crab chemosensory ability and can cause mortality (Buchanan et al. 1970; Caldwell et al. 1978; Pearson et al. 1980). Their commercial and recreational value makes them relevant to stakeholders.

Like English sole and Dungeness crab, littleneck clams are sediment dwellers and have a high probability of exposure to sediment-bound contaminants. Large numbers are known to occur in the study area (WDFW 1998) and are heavily harvested by both recreational and commercial clam diggers (WDFW 2002b). Because adult clams are sedentary, any response they exhibit is likely due to local stressors, providing a good indication of the local condition of the CP region.

Great blue heron use both intertidal and terrestrial habitats, providing a link between the aquatic and terrestrial components of the study area. Two large nesting colonies, consisting of about 300 nesting pairs each, are located within the study area at Point Roberts and Birch Bay (Kelsall 1989; Butler 1995; Fissinger 1996, 1998). Because great blue heron are predators and potentially prey on Pacific herring, English sole, and shellfish, they are likely to be exposed to and bioaccumulate persistent chemicals that may occur in the study area. Tissues from great blue heron in a nearby colony on Boundary Bay were found to contain measurable quantities of PCBs, mercury, and DDE, a toxic metabolite of the pesticide DDT, which is known to cause eggshell thinning and mortality in birds (Elliot et al. 1989; Bellward et al. 1990; Hart et al. 1991). Finally, because the local subspecies (*Ardea herodius fannini*) is nonmigratory (Butler 1995), great blue heron provide an indication of the local condition of the CP region, reducing the probability of observing effects caused by stressors outside the region.

These assessment endpoints were chosen because they are valued by stakeholders and are ecologically important. Each endpoint is important to different stakeholders in the region, ranging from members of the commercial fishing fleets to recreational users of beaches for clam digging or bird watching. Each species is also on WDFW's Priority Habitats and Species list (WDFW 2002c), illustrating their value as ecological resources of the state of Washington. They are known to occur in and utilize the study area, have a high probability of exposure to potential stressors in the region, and utilize different components of the nearshore ecosystem.

Identification of Habitats

Habitats were identified according to the classification system defined by WDNR's Nearshore Habitat Program (WDNR 1997b), USEPA Region 10 Estuarine

Habitat Assessment Protocol (Simenstad et al. 1991), and the published literature about the habitat requirements of the chosen assessment endpoints (Pauley et al. 1986; Toole et al. 1987; McMillan 1991; Alexander et al. 1993; Butler 1995; Pentilla 1997; Hirose and Kawaguchi 1998). The ten habitats represent different vegetation and substrate types in the upland, intertidal, and subtidal areas in the study area. The ten habitats are: (1) gravel cobble intertidal, (2) sandy intertidal, (3) nearshore soft bottom subtidal, (4) intertidal mudflats, (5) inter- or subtidal eelgrass, (6) inter- or subtidal macroalgae, (7) water column, (8) stream, (9) wetlands, and (10) forest.

Identification of Sources of Stressors

The nearshore lands in the CP region are dominated by agriculture interspersed with residential, industrial, forested, and undeveloped lands. Large shipping vessels travel to and from three deepwater shipping piers (Figure 13.1), and hundreds of recreational and fishing vessels have moorage in private and public marinas in the area (WDNR 1997a). Beaches are popular for clam digging, crabbing, and other recreational uses. To portray this mixture of multiple human uses, we partitioned anthropogenic sources of stressors into eight categories for use in the RRM: (1) accidental spills, chemical spills, (2) agricultural landuse, (3) ballast water, (4) piers, (5) point sources of pollution, (6) recreational activities, (7) urban landuse, and (8) vessel traffic. Natural sources of stressors were eliminated from this study due to a lack of site-specific data and in order to limit the study to sources relevant to the regional landuse, nearshore, and coastal management decisions facing local managers.

Accidental spills occur in both terrestrial and aquatic habitats in the study area. Chemicals released as accidental spills that were reported to the Washington Department of Ecology since 1995 included petroleum, both crude oil and fuel spills, automotive chemicals such as antifreeze, and pesticides and herbicides (Ecology 2001). Spills ranged in volume from less than a pint to hundreds of gallons. While most spills in the ecology database released only small amounts of contaminants, the cumulative effects of many small spills have the potential to cause direct or indirect effects to terrestrial and aquatic biota in the region.

Agricultural landuse also introduces stressors into the CP environment. Agriculture dominates the upland landscape, occupying 41% of the land in the study area (Whatcom County Assessor 2000). Runoff from agricultural land increases nutrient levels, siltation, and turbidity in streams and offshore waters. Pesticides and herbicides can run off and enter surface water and potentially cause toxicity. Removal of natural terrestrial habitat for agriculture can have direct and indirect effects on valued species.

Ballast water released from large shipping vessels can contain contaminants and introduce exotic species into the marine waters. If they become established, these introduced species have the potential to cause physical and behavioral disturbances to native organisms, out-competing them for food, space, or other valuable resources.

Large piers act as a source of stressors by changing nearshore sediment drift patterns, causing beach starvation in some areas, and enhancing sediment deposition in others (MacDonald et al. 1994). Piers may also shade out nearshore vegetation

(MacDonald et al. 1994) and introduce contaminants from pilings treated with antifouling agents.

Stack and effluent emissions, defined in this assessment as point source pollution, introduce chemical stressors into the environment. While the National Pollution Discharge and Elimination System (NPDES) permits regulate most emissions in the area, small amounts of contaminants may potentially cause toxicity to organisms in the nearshore environment.

Recreational activities, including clam digging and crabbing, affect organisms in the environment in a number of ways and occur extensively in some parts of the study area (WDFW 2001a). Clam digging has the obvious effect of mortality by harvesting animals, but the presence of humans in habitats might also cause behavioral disturbances to other species. Clam digging can also change small-scale habitat composition by removing cobbles, exposing the gravel and sand matrix which is easily erodable by wave action and killing other organisms. Holes left by the removal of cobbles create tide pools laden with sediment and decaying organic matter, which may reduce the amount of available habitat for native species (Kyte 2001).

Urban and industrial landuse make up 19% of the total landuse in the CP study area (Whatcom County Assessor 2000). Runoff from streets, yards, and parking lots can, like agricultural runoff, result in increased nutrients, siltation, and turbidity in streams and nearshore marine waters. Traffic and other noises can disturb wildlife in adjacent forest or nearshore habitats.

Finally, commercial and recreational vessel traffic in the region can cause behavioral disturbances to fish and wildlife, introduce exotic species, increase turbidity of nearshore waters, and introduce contaminants through fuel leaks and antifouling agents.

Conceptual Model Development

We developed a conceptual model (Figure 13.3) to depict the interconnections among sources, stressors, habitats, and endpoints based on information in the published and unpublished literature (Pauley et al. 1986; Simenstad et al. 1991; Alexander et al. 1993; Thom and Shreffler 1994; EVS 1999). The conceptual model depicts preliminary exposure and effects filters for each source–stressor–habitat–endpoint combination. A complete exposure pathway met the following criteria based on a review of published and unpublished literature: the source releases or causes the stressor, the stressor will occur and persist in the habitat, the endpoint uses the habitat type, the stressor can negatively affect the assessment endpoint.

The problem formulation process resulted in maps and a conceptual model that later became the foundation of the analysis and risk characterization phases of the assessment.

RISK ASSESSMENT METHODS

The Cherry Point RRM risk assessment process followed in general the USEPA guidelines (1998). Accordingly, the problem formulation phase of the assessment

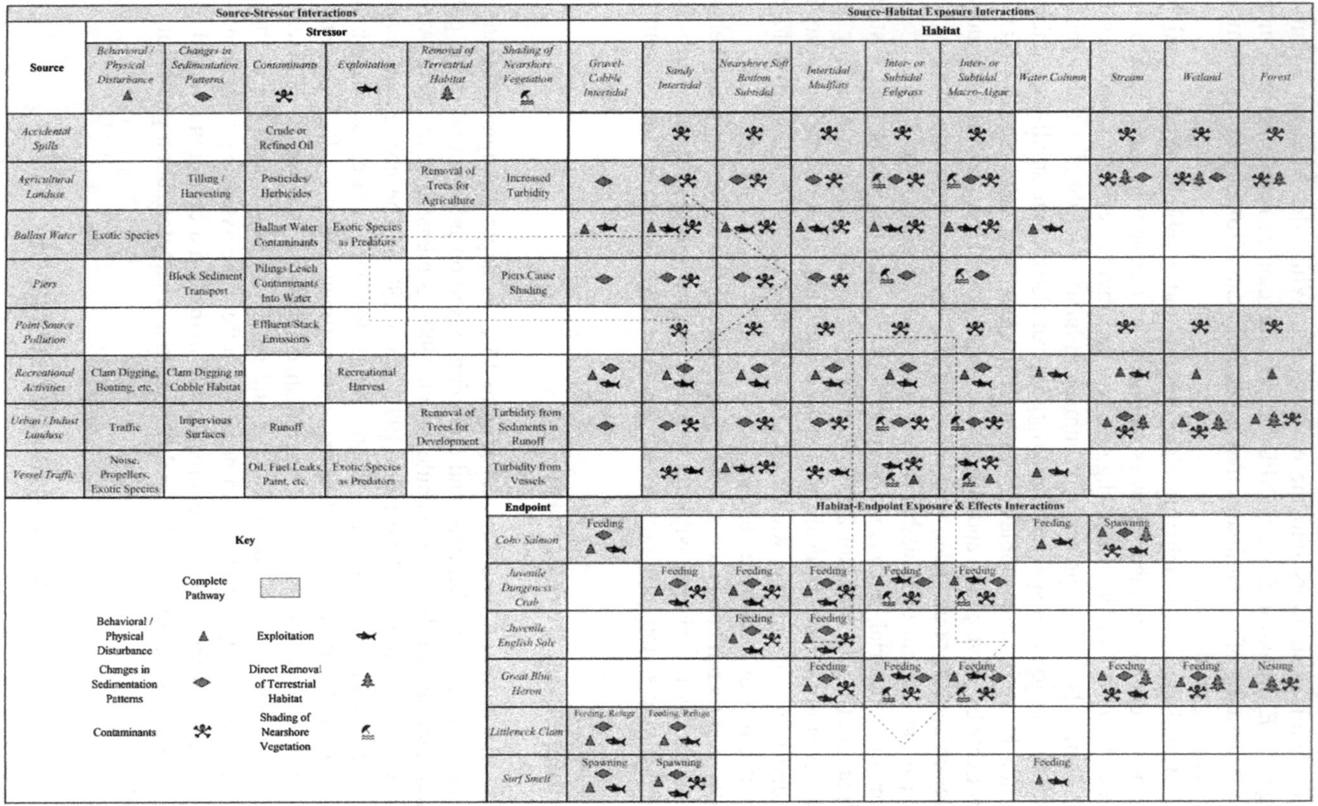

Figure 13.3 Conceptual model depicting potential exposure and effects pathways from source to stressor to habitat to endpoint.

led into analysis, risk characterization, and uncertainty analysis. Analysis and risk calculation methods were similar to those used in previous RRM regional risk assessments (Landis and Wiegers 1997; Wiegers et al. 1998; Landis et al. 2000b; Walker et al. 2001; Chen 2002; Obery and Landis 2002; Moraes et al. 2002). Risk characterization was founded on the following assumptions (Landis and Wiegers 1997; Wiegers et al. 1998):

- The greater the size or frequency of a source in a subregion, the greater the potential for exposure to stressors
- The type and density of assessment endpoints are related to the available habitat
- The sensitivity of receptors to stressors varies between habitats
- The severity of effects in subregions of the CP region depends on relative exposures and the characteristics of the organisms present

Analysis

According to the assumptions above, GIS and other site-specific data (Table 13.1) were used to rank (1) sources of stressors (e.g., human landuse, point sources of pollution, vessel traffic) and (2) habitats (e.g., cobble–gravel intertidal habitat, wetlands) for subregions within the study area. We then assigned exposure and effects filters for each source–stressor–habitat endpoint combination based on the conceptual model (Figure 13.3) as well as geographic data. Ranks and filters were integrated to derive risk estimates for subregions, sources, and endpoints in the study area.

Sources and Habitat Ranks

Geographical datasets were used to assign source and habitat ranks to the six subregions in the study area. Using Jenk's optimization in ArcView™ GIS, datasets from Table 13.1 were broken into four categories to assign ranks of 0, 2, 4, and 6. Table 13.2 and Table 13.3 provide the criteria for the source- and habitat-ranking schemes. Table 13.4 and Table 13.5 contain the risk ranks assigned for each source and habitat for subregions.

Exposure and Effects Filters

Exposure and effects filters of 0, 0.5, or 1 were assigned to reflect low, medium, or high probability of exposure or effects for each source to endpoint combination. These filters were based primarily on linkages described in the conceptual model (Figure 13.3).

Exposure filters received a score of 1 if the conceptual model pathway between source and habitat was complete and a 0 if the pathway was not complete. A score of 1 was reduced to 0.5 if site-specific data indicated that the stressor occurred in small amounts, thus reducing the probability of exposure to endpoints. Site-specific data were only available for one stressor — contaminants. Exposure filters for sources releasing contaminants into Point Roberts and Birch Bay were changed to 0.5 because sediments collected from these subregions caused low or no toxicity in

Table 13.1   Geographical Information Used in the Cherry Point RRM

Name	Data Description	Data Source
Accidental spills	Locations and volumes of spills ranging from 1 pint to hundreds of gallons	Ecology (2001)
Landuse	Landuse as designated by the Whatcom County Assessor's 2000 tax assessment codes	Whatcom County Assessor (2000); Whatcom County PUD (2000)
Ballast water releases	Locations, dates, and volumes of ballast water releases from 1999–2001	WDFW (2001b)
Piers	Locations of piers and docks on Washington coasts	WDNR (1997a)
Point sources of pollution	Locations of NPDES permit holders, toxic release inventory sites, and solid and hazardous waste sites	USEPA (2001)
Recreational clam diggers and crab buoy locations	Locations of Washington Department of Fish and Wildlife aerial observations of recreational shellfish harvesters and crabbers 2001	WDFW (2001a)
Vessel traffic	Locations of boat slips for both recreational and commercial vessels; Washington Department of Natural Resources Shorezone Inventory	WDNR (1997a)
Intertidal substrates and vegetation	Locations of nearshore habitat types	WDNR (1997b)
Bathymetry	Sea floor depths as measured by National Ocean Service	NOS (2001)
Streams	Streams and rivers	USGS (1987)
Wetlands	Location and area of wetlands	Whatcom County Planning and Development Services (1998)
Forest	Land parcels designated as forest based on Whatcom County Assessor's 2000 tax assessment codes	Whatcom County Assessor (2000); Whatcom County PUD (2000)

1999 (Ecology 1999). Drayton Harbor filters retained their value of 1 because sediments from this region caused toxicity in several tests (Ecology 1999). CP contaminant filters received filter values of 1 because waters in this region are listed on Washington Department of Ecology's Section 303(d) list of impaired water bodies due to sediment contamination (Ecology 1998).

Likewise, an effects filter received a score of 1 if the conceptual model pathway from habitat to endpoint was complete and a score of 0 for an incomplete pathway. A score of 1 was reduced to 0.5 if site-specific data indicated the endpoint uses the habitat only marginally, reducing the probability of exposure, and therefore effects. If no site-specific data were available, the score was left as 1.

Effects filters were also assigned according to the conceptual model and site-specific data. Site-specific data were available for great blue heron, surf smelt embryos, and juvenile Dungeness crab. Great blue heron effects filters were changed according to average foraging densities from aerial bird counts from 1992 to 1999

Table 13.2   Criteria for Ranking Sources of Stressors

Source	Ranking Criterion	Range (divided by natural breaks)*	Rank	Drayton Harbor Example
Accidental spills	Volume of spills (gallons) per year per km² in subregions (Ecology, unpublished data)	0 0.001–0.023 0.024–35.395 35.396–280.677	0 (zero) 2 (low) 4 (medium) 6 (high)	255.8 gallons = rank of 6
Agricultural land use	Percent agricultural land (Whatcom County Assessor 2000; Whatcom County PUD 2000)	0 0.69–16.94 16.95–42.02 42.03–50.41	0 (zero) 2 (low) 4 (medium) 6 (high)	42.02% agriculture = rank of 4
Ballast water	Ballast water released? (WDFW, unpublished data)	No — — Yes	0 (zero) 2 (low) 4 (medium) 6 (high)	No ballast water release = rank of 0
Piers	Number of large piers or docks per km shoreline (WDNR 1997a)	0 0.001–0.032 0.033–0.074 0.075–0.243	0 (zero) 2 (low) 4 (medium) 6 (high)	0.0319 piers per km shoreline = rank of 2
Point sources of pollution	Number of point sources of pollution in region per km² land (USEPA 2001)	0 0.001–0.04 0.050 0.18 0.19–0.020	0 (zero) 2 (low) 4 (medium) 6 (high)	0.18 point sources per km² land = rank of 4
Recreational activities	Number of recreational clam diggers per km shoreline (WDFW 2001a)	0 0.001–8.733 8.734–21.762 21.763–156.127	0 (zero) 2 (low) 4 (medium) 6 (high)	7.9 recreational clam diggers per km shoreline = rank of 2
Urban and industrial landuse	Percent urban land (Whatcom County Assessor 2000; Whatcom County PUD 2000)	0 0.01–21.00 21.01–31.17 31.18–41.34	0 (zero) 2 (low) 4 (medium) 6 (high)	28.3% urban land = rank of 4
Commercial/ recreational vessel traffic	Number of slips per km shoreline (plus areas with known vessel traffic) (WDNR 1997a)	0 0.001–11.64 11.65–29.878 29.879–51.025	0 (zero) 2 (low) 4 (medium) 6 (high)	29.878 boat slips per km shoreline = rank of 4

* Significant figures to three decimal places

(WDFW 1999) and locations of large nesting colonies. Using these criteria, great blue heron filters for Point Roberts, Drayton Harbor, Birch Bay, and Lummi Bay kept filters of 1 and Cherry Point and Alden Bank filters were changed to 0.5.

Surf smelt embryo filters were adjusted based on the amount of spawning habitat on the beaches in the region (WDFW 2001c) divided by natural breaks using Jenk's optimization in ArcView™ GIS. Subregions with between 1 and 2 km of spawning beach received filter values of 0.5. Filters for subregions with between 4 and 8 km retained their filter values of 1.

Finally, we adjusted juvenile Dungeness crab filters according to the shoreline characteristics in the subregions. Juvenile Dungeness crab are found in higher densities

Table 13.3   Criteria for Ranking Habitats

Habitat	Ranking Criteria	Range (divided by natural breaks)*	Rank	Drayton Harbor Example
Gravel-cobble intertidal (WDNR 1997b)	Area (km²)	0 0.062–0.271 0.272–0.636 0.636–2.316	0 (zero) 2 (low) 4 (medium) 6 (high)	2.316 km² = rank of 6
Sandy Intertidal (WDNR 1997b)	Area (km²)	0 0.001–0.852 0.853–1.894 1.895–8.914	0 (zero) 2 (low) 4 (medium) 6 (high)	1.783 km² = rank of 4
Mudflats (WDNR 1997b)	Area (km²)	0 0.009 — 0.347	0 (zero) 2 (low) 4 (medium) 6 (high)	0 km² = rank of 0
Eelgrass (WDNR 1997b)	Area (km²)	0 0.245–1.367 1.368–3.755 6.491–6.922	0 (zero) 2 (low) 4 (medium) 6 (high)	6.493 km²= rank of 6
Macroalgae (WDNR 1997b)	Area (km²)	0 0.052–0.238 0.239–0.976 0.977–1.212	0 (zero) 2 (low) 4 (medium) 6 (high)	0.052 km² = rank of 2
Subtidal soft substrate (NOS 2001)	Area (km²)	0 16.026–33.455 33.455–60.122 60.122–94.196	0 (zero) 2 (low) 4 (medium) 6 (high)	16.026 km² = rank of 4
Water column (NOS 2001)	Area (km²)	0 16.026–33.455 33.455–82.099 82.099–145.754	0 (zero) 2 (low) 4 (medium) 6 (high)	16.026 km² = rank of 2
Streams (USGS 1987)	Length (km)	0 0.001–7.015 7.016–42.734 42.735–159.984	0 (zero) 2 (low) 4 (medium) 6 (high)	109.683 km = RRM1 rank of 6
Wetlands (Whatcom County Planning and Development Services 1998)	Area (km²)	0 0.589–5.163 5.164–6.827 16.336–20.046	0 (zero) 2 (low) 4 (medium) 6 (high)	20.046 km² = rank of 6
Forest (Whatcom County Assessor 2000; Whatcom County PUD 2000)	Area (km²)	0 0.001–0.196 0.197–9.950 9.951–14.030	0 (zero) 2 (low) 4 (medium) 6 (high)	9.950 km² = rank of 6

* Significant figures to three decimal places

in protected bays (Gunderson et al. 1990; McMillan 1991). Subregions containing protected bays kept their filter values of 1, and filters for subregions with no protected bays were changed to 0.5.

Table 13.4 Habitat Ranks for Subregions

Habitat		Risk Region					
		Point Roberts	Drayton Harbor	Birch Bay	Cherry Point	Lummi Bay	Alden Bank
Gravel-cobble intertidal	RRM Rank	4	6	2	4	4	0
	(km²)	(0.636)	(2.316)	(0.062)	(0.436)	(0.272)	0.000
Sandy intertidal	RRM Rank	2	4	4	2	6	0
	(km²)	(0.609)	(1.783)	(1.894)	(0.853)	(8.914)	0.000
Mudflats	RRM Rank	2	0	0	0	6	0
	(km²)	(0.009)	(0.000)	(0.000)	(0.000)	(0.347)	(0.000)
Eelgrass	RRM Rank	4	6	4	2	6	0
	(km²)	(2.368)	(6.491)	(3.755)	(0.245)	(6.922)	(0.000)
Macroalgae	RRM Rank	4	2	6	4	2	0
	(km²)	(0.977)	(0.052)	(1.212)	(0.870)	(0.238)	(0.000)
Soft bottom subtidal	RRM Rank	6	4	2	2	2	6
	(km²)	(81.22)	(16.026)	(60.112)	(25.26)	(33.455)	(81.22)
Water column	RRM Rank	4	2	4	2	2	6
	(km²)	(82.099)	(16.026)	(60.112)	(25.260)	(33.455)	(145.754)
Stream	RRM Rank	0	6	4	2	6	0
	(km)	(0.000)	(109.683)	(42.734)	(7.015)	(159.984)	(0.000)
Wetland	RRM Rank	2	6	6	4	4	0
	(km²)	(0.589)	(20.046)	(16.336)	(5.164)	(6.827)	(0.000)
Forest	RRM Rank	2	6	4	2	6	0
	(km²)	(0.196)	(9.95)	(4.34)	(0.12)	(14.03)	(0.000)

Data from Table 13.3 used to make the assigned ranks are in parentheses.

Table 13.5 Source Ranks for Subregions

Source		Risk Region					
		Point Roberts	Drayton Harbor	Birch Bay	Cherry Point	Lummi Bay	Alden Bank
Accidental spills	RRM rank (gallons/km²)	2 (0ᵃ)	6 (255.84)	2 (0.02)	4 (35.39)	6 (280.68)	0 (0)
Agricultural landuse	RRM rank (% landuse)	2 (1.39)	4 (42.02)	4 (35.44)	2 (16.94)	6 (50.41)	0 (0)
Ballast water	RRM rank (yes/no)	0 (no)	0 (no)	0 (no)	6 (yes)	0 (no)	0 (no)
Piers	RRM rank (#/km shoreline)	4 (0.07)	2 (0.03)	0 (0.00)	6 (0.24)	2 (0.03)	0 (0)
Point source pollution	RRM rank (#/km²)	0 (0.00)	4 (0.18)	2 (0.04)	6 (0.20)	4 (0.16)	0 (0)
Recreational activities	RRM rank (# ind./km shoreline)	0 (0)	2 (7.98)	6 (156.13)	4 (21.76)	2 (8.73)	0 (0)
Urban and industrial landuse	RRM rank (% landuse)	4 (31.17)	4 (28.30)	4 (29.04)	6 (41.34)	2 (21.00)	0 (0)
Vessel traffic	RRM rank (# slips/km shoreline)	6 (51.03)	4 (29.88)	2 (11.64)	6 (0.081ᵇ)	6 (5.694ᵇ)	4 (0ᵇ)

ᵃ Point Roberts received a rank of 2 instead of 0 due to high uncertainty associated with the Ecology spills database (Ecology 2001), which documented spills but recorded no volume.

ᵇ Cherry Point, Lummi Bay, and Alden Bank ranks were increased to 6, 6, and 4, respectively, due to a high amount of vessel traffic in these subregions, despite low slip density (WDNR 1997a).

Data from Table 13.3 used to make the assigned ranks are in parentheses.

Risk Characterization

We integrated source and habitat ranks with exposure and effects filters to determine the relative risk estimates. Risk estimates were derived by first multiplying the source and habitat ranks by the exposure and effects filters for each subregion. The sum of the products of each source–habitat filter combination determined the final estimate of risk. These risk estimates were compared among subregions, sources, habitats, and endpoints to reveal:

1. The subregions where most risk occurs.
2. The sources contributing the most risk.
3. The habitats where most risk occurs.
4. The endpoints most at risk in the Cherry Point area.

Uncertainty Analysis

Uncertainty analysis differed from previous RRM assessments with the addition of an alternative habitat ranking scheme to analyze the effects of model uncertainty and Monte Carlo techniques to quantitatively describe parameter uncertainty in risk predictions (Warren-Hicks and Moore 1998). The risk predictions produced in the RRM are point estimates based on ranks and filters derived from imperfect data. To communicate the uncertainty associated with these point estimates, Monte Carlo analysis was used to generate distributions of probable predictions for each risk component. In addition to using Monte Carlo analysis to describe parameter uncertainty in the assessment, we also applied an alternative habitat ranking scheme to the RRM to investigate uncertainty in the model and the effects of habitat ranking assumptions on the risk estimates.

Monte Carlo Analysis

The first phase of uncertainty analysis applied Monte Carlo techniques to analyze parameter uncertainty in the risk predictions. In risk assessment, Monte Carlo uncertainty analysis combines assigned probability distributions of input variables to estimate a probability distribution for output variables (Burmaster and Anderson 1994). In the case of the Cherry Point regional risk assessment, the input variables are the ranks and filters with medium or high uncertainty and the output variables are the risk estimates.

For the Monte Carlo uncertainty analysis, we first assigned designations of low, medium, or high uncertainty to each source, habitat rank, exposure, and effects filter based on data quality and availability. We assigned discrete probability distributions to ranks and filters with medium and high uncertainty according to the criteria in Table 13.6 and Table 13.7. We did not assign distributions to ranks and filters with low uncertainty but left them simply as the original point estimate.

We assigned high uncertainty to accidental spills ranks for all subregions because the Ecology (2001) spills dataset was incomplete, lacked spill volume for many of the records, and locations of several records in the database were undeterminable. This poor data quality resulted in high uncertainty in the accidental spills ranks.

Table 13.6 Uncertainty Analysis Monte Carlo Input Distributions for Ranks with Medium and High Uncertainty

Assigned Rank Value	Uncertainty	Assigned Probability (%) for Ranks			
		0	2	4	6
0	High	60	20	20	0
0	Medium	80	10	10	0
2	High	0	60	20	20
2	Medium	0	80	10	10
4	High	0	20	60	20
4	Medium	0	10	80	10
6	High	0	20	20	60
6	Medium	0	10	10	80

Table 13.7 Uncertainty Analysis Monte Carlo Input Distributions for Filters with Medium and High Uncertainty

Assigned Filter Value	Uncertainty	Assigned Probability (%) for Ranks		
		0	0.5	1
0	High	60	20	20
0	Medium	80	10	10
0.5	High	0	60	40
0.5	Medium	0	80	20
1	High	0	40	60
1	Medium	0	20	80

The recreational activities rank for Point Roberts was assigned high uncertainty because the WDFW dataset (2001a) on which ranks were based did not survey the Point Roberts region. We assigned gravel-cobble, sandy intertidal, and mudflats habitat ranks for the Point Roberts subregion medium uncertainty because no habitat area data were available for this region and had to be derived using GIS analysis and alternative datasets (WDNR 1997a; NOS 2001).

Eelgrass and macroalgae habitat ranks for the Alden Bank subregion were assigned high uncertainty because no vegetation data were available this far off shore. These habitat ranks received ranks of 0; however, because the sea floor depth becomes quite shallow again at Alden Bank (NOS 2001), some vegetation is most likely present; the amount is undetermined.

We assigned medium uncertainty to subtidal soft bottom habitat ranks for all subregions because subtidal substrate data were unavailable for the entire study area. We derived areas for this habitat on which ranks were based using GIS analysis and bathymetry data (NOS 2001), assuming the majority of subtidal substrate to be soft bottom (vs. vegetation or rocky substrate). While this assumption may overestimate the amount of soft substrate on the sea floor bottom, it overestimates this habitat in all subregions evenly. While the area values are not precise, the final ranks most likely represent the relative amount of this habitat type in each subregion and are therefore appropriately assigned, albeit with a degree of uncertainty.

Inconsistent landuse data quality resulted in medium uncertainty in the Lummi Bay forest habitat rank. Because a large portion of the Lummi Bay subregion falls within the boundaries of the Lummi Nation Indian reservation, the Whatcom County Assessor's tax parcel dataset (Whatcom County Assessor 2000) did not accurately cover these areas. Instead, landuse areas in this subregion were assigned according to another dataset developed by the Whatcom County Public Utility District (Whatcom County PUD 2000). Inconsistency of the datasets warranted assigning medium uncertainty to the forest habitat rank in the Lummi Bay subregion.

Vessel traffic ranks for Lummi Bay, Cherry Point, and Alden Bank were assigned medium uncertainty because the dataset on which we based ranks for the other subregions (number of slips per km shoreline) did not accurately portray the amount of vessel traffic occurring as a result of the industrial piers and ferry terminal in these subregions. Taking this into account but lacking a dataset that characterized both recreational and commercial vessel traffic, we instead assigned ranks of the next higher category to these subregions.

All other ranks were assigned low uncertainty. Filters similarly received designations of high, medium, or low uncertainty. A lack of understanding of the fate and transport of stressors, deficient site-specific information about the locations and amounts of stressors, and variance in the quantity of a stressor that sources may release were all grounds for assigning medium and high uncertainties to filters.

Using Crystal Ball® 2000 software as a macro in Microsoft® Excel 2002, we ran the Monte Carlo simulations for 1000 iterations and derived output distributions for each subregion, source, habitat, and endpoint risk prediction. These distributions show a range of probable risk estimates associated with each point estimate.

Alternative Habitat Ranking Scheme

During the second phase of the uncertainty analysis, we applied an alternative habitat ranking method to investigate the effects of the underlying assumptions of the habitat ranking scheme on the final risk estimates. The original RRM method assumes that a large amount of habitat in a subregion increases the probability that an organism utilizing that habitat will come into contact with a stressor, thus increasing the probability of exposure and, therefore, risk. Accordingly, subregions with a larger amount of habitat receive a high rank, signifying a high probability of impact to endpoints. This ranking method becomes problematic when analyzing risk at a population, rather than an individual or organism scale, because the effects of stressors differ between individuals and populations.

The alternative method for ranking habitat in subregions has an entirely different set of assumptions. This second method assigns high ranks to subregions with a small amount of habitat and assumes:

- A small habitat size supports a small population of organisms. This small population would theoretically be more susceptible to the effects of stressors in the environment and is at greater risk of becoming extinct than a larger, more resilient, population.
- Stressor concentration is greater in small habitats, thus increasing the likelihood of both exposure and effects.

Table 13.8   Risk Scores for the Risk Regions

	Risk Region						
Habitat	Point Roberts	Drayton Harbor	Birch Bay	Cherry Point	Lummi Bay	Alden Bank	Total
Gravel-cobble intertidal	192	300	168	528	220	0	1408
Sandy intertidal	125	440	420	400	612	0	1997
Mudflats	125	0	0	0	756	0	881
Eelgrass	282	744	376	212	696	0	2310
Macroalgae	282	248	564	424	232	0	1750
Soft bottom subtidal	279	368	124	246	172	180	1369
Water column	96	36	128	128	64	48	500
Stream	0	504	304	144	480	0	1432
Wetland	40	240	192	88	152	0	712
Forest	28	192	96	36	180	0	532
Total	1449	3072	2372	2206	3564	228	12,891

We investigated the uncertainty associated with these assumptions by applying the alternative habitat ranking scheme to the RRM and compared these results with the original RRM risk predictions. We also performed Monte Carlo uncertainty analysis on the alternative habitat rank results.

RESULTS

Risk Characterization

Risk calculations (Table 13.8) revealed (1) the subregions and (2) habitats where most of the risk occurs, (3) which sources contribute the most risk, and (4) the endpoints most likely to be affected by anthropogenic stressors in the Cherry Point region. The risk predictions resulting from this assessment are estimates about the relative risk to endpoints in the region. These patterns of risk form hypotheses that can be tested.

The RRM predicted the highest risk in Lummi Bay and Drayton Harbor, medium risk in Cherry Point, Birch Bay, and Point Roberts, and low risk in Alden Bank (Figure 13.4). Habitats where most risk occurs are eelgrass, sandy intertidal, and macroalgae (Table 13.8). The major contributors of risk in the region are commercial and recreational vessel traffic, upland urban and agricultural landuse, and shoreline recreational activities (Figure 13.5). The biological endpoints most likely to be at risk are great blue heron and juvenile Dungeness crab (Figure 13.6).

Vessel traffic was identified as a major contributor of risk in Point Roberts, Drayton Harbor, Lummi Bay, and Alden Bank subregions. Urban landuse was important in Drayton Harbor and Cherry Point. The model predicted that agricultural landuse contributed much of the risk in Drayton Harbor and Lummi Bay. Recreational activities were important in Birch Bay. Ballast water was the most important source for Cherry Point. All other sources ranked comparatively low (Figure 13.5).

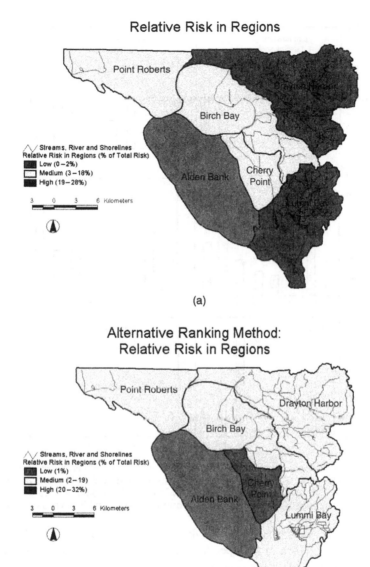

Relative Risk in Regions

(a)

**Alternative Ranking Method:
Relative Risk in Regions**

(b)

Figure 13.4 Comparisons of the relative risks depending upon assumptions about the sensitivity of the habitat vs. area. (See color insert following page 178.)

Uncertainty Analysis

The model developed for this risk assessment was based on a combination of site-specific data and general knowledge about interconnections between risk components.

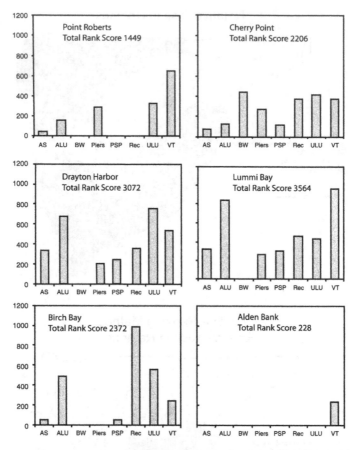

Figure 13.5   Relative contribution to risk from sources in subregions. Y-axis is the relative
risk score; X-axis from left to right: AS = accidental spills, ALU = agricultural
landuse, BW = ballast water, Piers = piers, PSP = point source pollution, Rec
= recreational activities, ULU = urban landuse, and VT = vessel traffic.

Uncertainty in the assessment arose from flaws in input data and imperfections in the
model and include a lack of site-specific data in some or all subregions within the
study area, poor data quality, misunderstanding of the fate and transport of stressors
in the CP environment, omitting contributing sources and stressors, a failure to
identify and incorporate temporal and spatial patterns, and from incorrect assumptions in the model.

To quantify the effects of parameter uncertainty on the risk predictions, Monte
Carlo analysis was applied to the RRM to derive probability distributions of possible
risk estimates. The Monte Carlo analysis resulted in probability distributions of risk
predictions for each subregion, habitat, source, and assessment endpoint. Results for
subregions, sources, and endpoints are depicted in Figure 13.7.

During the second component of uncertainty analysis, we applied an alternative
habitat ranking scheme to investigate the assumptions about habitat use by biological
assessment endpoints and how those assumptions affect the predicted risk values.

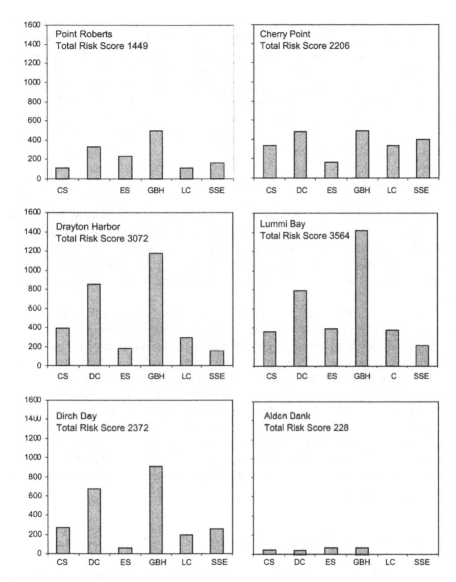

Figure 13.6 Relative risk to biological assessment endpoints in subregions. Y-axis is the relative risk score; X-axis from left to right: CS = Coho salmon, DC = juvenile Dungeness crab, ES = juvenile English sole, GBH = great blue heron, LC = native littleneck clam, and SSE = surf smelt embryos.

Monte Carlo Analysis

The Monte Carlo analysis produced approximately normal distributions with means close to the predicted risk values for most risk predictions, suggesting low uncertainty in RRM predictions for most risk components.

The Alden Bank Monte Carlo distribution was the narrowest of all the distributions for subregions, suggesting the most confidence in the risk prediction; Point

Roberts, Lummi Bay, and Cherry Point had the widest distributions, indicating less certainty in these predictions. Point Roberts and Alden Bank distributions were right-skewed, suggesting the possibility that, despite the mean and risk prediction being similar, the models may have underestimated risk in these subregions. The Lummi Bay distribution was left-skewed, suggesting the possibility of an overestimation of risk (Figure 13.7). For source contribution to risk, the RRM prediction for vessel traffic was on the upper end of the Monte Carlo distribution of risk values. The distribution was also left-skewed. Taken together, these indicate the model might have overestimated the contribution of risk from vessel traffic. The distribution for recreational activities was right-skewed, suggesting a possible underestimation of risk from this source. The ranges of the distributions for agricultural landuse, ballast water, piers, point sources of pollution, and urban landuse were relatively narrow, demonstrating high confidence in their risk predictions. Comparatively, the ranges of the Monte Carlo probability distributions for accidental spills, recreational activities, and vessel traffic were wider, demonstrating less confidence in their predictions (Figure 13.7). These results are consistent with the relatively poor data quality and higher uncertainty in the initial ranks for accidental spills, recreational activities, and vessel traffic.

Monte Carlo distributions for habitats also revealed components with high and low uncertainty. The ranges of the distributions for mudflats, streams, and wetlands were narrower than the others, suggesting higher confidence in habitat risk predictions for these three habitats than for the remaining seven. Distributions for mudflats and streams were left-skewed, however, suggesting a possible overestimation of risk. Eelgrass, macroalgae, and soft bottom subtidal distributions were right-skewed, indicating the model may have underestimated risk in these habitats.

Probability distributions for biological endpoints were indicative of an overestimation of risk in the predictions for juvenile Dungeness crab and surf smelt embryos, as evidenced by their right-skewed probability distributions. All other biological assessment endpoints had approximately normal distributions and low uncertainty (Figure 13.7).

Alternative Habitat Ranking Scheme

The alternative habitat ranking scheme did not change risk predictions for assessment endpoints, habitats, and sources, suggesting the RRM is fairly robust to changes in habitat ranks. However, the spatial distribution of risk changed when the alternative habitat ranking scheme was applied. Risk in Lummi Bay and Drayton Harbor moved from high risk in the original assessment to medium risk using the alternative ranking scheme. The CP region moved from its original medium risk to high risk (Figure 13.4b). The Monte Carlo uncertainty results from the alternative habitat ranking scheme calculations were similar to the original CP RRM Monte Carlo results.

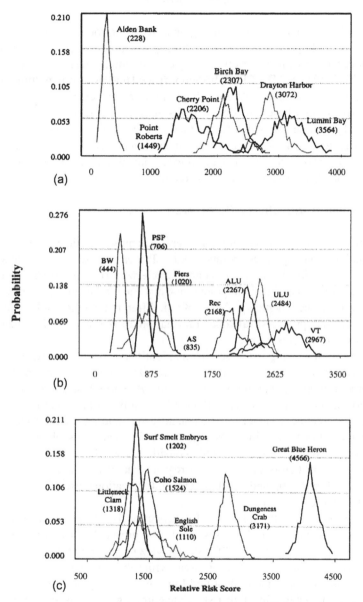

Figure 13.7   Monte Carlo results of the RRM calculations (with RRM risk scores in parentheses) for (a) subregions, (b) sources of stressors, and (c) biological endpoints. AS = accidental spills; ALU = agricultural landuse; BW = ballast water;
PSP = point source pollution; Rec = recreational activities; ULU = urban
landuse; and VT = vessel traffic. Lines do not depict continuous distributions
but are the outlines of discreet distributions.

DISCUSSION

The need to characterize risk at regional scales introduces opportunities and challenges to ecological risk assessors. Multiple sources, stressors, habitats, and endpoints can interact in a variety of ways, confounding results and introducing error in risk calculations. The CP study demonstrated that risk from multiple stressors to varying biological endpoints in a landscape can be characterized using this spatially explicit rank-based method. The results from the regional risk assessment are testable hypotheses on which future field and laboratory research can be based. Finally, 'the regional risk assessment can help environmental decision makers to plan not only future research, but to make present-day decisions about development, landuse, and shoreline use to reduce risk to biological endpoints in the study region.

Relative Risk in the Cherry Point Area

The first objective of this regional risk assessment was to determine the relative risks to multiple biological endpoints in the CP area for use in land management decisions. This regional risk assessment characterized risk in the CP region as it affects a variety of biological endpoints, not just Pacific herring as in past assessments. This assessment depicts a picture of risk different than that for Pacific herring in the region and provides valuable information to regional land managers not available from past assessments.

The CP Pacific herring risk assessments identified historical overexploitation as the major anthropogenic stressor in the region. Exploitation does not play a major role for other biological endpoint species in the CP region. Instead, the most important anthropogenic sources of stressors identified in this assessment included commercial and recreational vessel traffic, upland urban and agricultural landuse, and shoreline recreational activities. Managers making decisions about upland and aquatic landuses in the CP region should consider these additional sources of stressors as major contributors of risk in the region. Future decisions that control the effects of vessel traffic, upland development, and recreational use will help to reduce future risk in the CP region.

The locations where the majority of risk occurs also differ between the Pacific herring risk assessments and this assessment. The prospective RRM Pacific herring risk assessments (Landis et al. 2000b; Markiewicz et al. 2001) found little differentiation between subregions in the CP area, while the predictions and uncertainty analysis in this risk assessment identified Lummi Bay, Drayton Harbor, and Cherry Point as areas of potentially high risk. Concentrating research, conservation, and restoration efforts in these subregions will also help to both better understand and reduce future overall risk in the study area.

The risk predictions generated by the CP RRM can aid regional land managers in making decisions by pinpointing important risk factors in each sub-region. In this sense, the RRM risk results are superior to many measures of ecosystem quality (e.g., indices of biotic integrity [IBI], habitat suitability indices, risk predictions based on quotients) for use in a decision-making framework because the predictions are more than simple point estimates of risk. Risk estimates are specific for subregions within the study area (Figure 13.5 and Figure 13.6). Land managers can therefore cater risk management schemes to address important factors in each subregion. For example, recreational activities contributed most to risk in the Birch Bay subregion, while agricultural and urban landuse appear to be the driving forces in the Drayton Harbor and Lummi Bay subregions. Risk management plans for the region can focus on reduction of source inputs from agricultural and urban landuse in Drayton Harbor and Lummi Bay, and on controlling recreational clam harvests in Birch Bay. These individualized risk management plans make the best use of resources and provide the most cost-effective means for reducing risk in the entire CP region. The RRM is also superior to the use of an IBI for environmental decision making because, unlike the IBI, assessment endpoints are chosen by stakeholders. Interpretation of RRM results is tied to stakeholder values and not to predetermined endpoints chosen by assessors and unlinked to management decisions.

The RRM can also be used as a predictive tool for regional managers. By defining decision options and comparing how these options will change input data, the RRM can be used to predict how risk scores change under various management scenarios. Using relatively little time and resources, decision options can be weighed against each other to identify the most effective management scheme for reducing risk in the region.

Application of the Relative Risk Model to Cherry Point

The second objective of this assessment was to evaluate the applicability of the RRM method in the region. The RRM was successfully applied to the CP region with only a few revisions. Revisions to the model included the addition of an alternative habitat ranking method and the use of Monte Carlo techniques in the uncertainty analysis.

Monte Carlo Uncertainty Analysis Techniques

Monte Carlo techniques allowed a more complete characterization of uncertainty than previous RRM uncertainty analysis methods have provided. In addition to qualitatively describing the sources of uncertainty, this quantitative method allows risk managers to know the degree of uncertainty, the possibilities of over- or underestimating risk, and which parameters contribute most to that uncertainty. The results, however, depend on the assumptions about the input distributions used to derive Monte Carlo probability output distributions. The more accurate these input distributions, the more confidence in the final Monte Carlo output. Every effort was made in this study to remain consistent when assigning input rank and filter distributions in order to avoid misrepresenting uncertainty in the Monte Carlo analysis.

The probability distributions resulting from this analysis better communicate confidence in risk estimates than the qualitative means used in past assessments of

Cherry Point. Environmental managers can decide whether the degree of uncertainty for risk estimates is acceptable for the decisions at hand and concentrate their efforts and resources on studies aimed to reduce uncertainty where it will matter most — in the decision-making context. For example, the Monte Carlo analysis identified risk predictions for accidental spills, recreational activities, and vessel traffic as having a relatively high degree of uncertainty. Efforts to reduce the uncertainty in these estimates (e.g., more consistent record keeping in environmental spills data-bases) can reduce overall uncertainty in the CP RRM and improve the confidence in management decisions based on risk predictions.

Alternative Habitat Ranking Scheme

This RRM regional risk assessment applied an additional habitat ranking scheme to investigate the effects of underlying assumptions of the ranking method on final risk estimates. Each method has its strengths and weaknesses, and applying one method over the other depends on the availability of data and the question being asked. Future practitioners of the RRM must utilize information specific to their study area to decide on appropriate source and habitat ranking schemes.

To examine which habitat ranking method is most appropriate in predictions of risk in an RRM assessment, several factors must be examined. If the RRM is being applied in a region where no exposure data other than habitat quantity, quality, and location are available, the original ranking method may provide the most accurate depiction of risk. Using the alternative ranking scheme may underestimate risk to biological assessment endpoints in the entire study area because it would fail to characterize exposure altogether. However, relying on habitat measures as an indi-cation of exposure may underestimate risk in subregions containing small amounts of habitat, where small populations may be exposed to stressors at a high concen-tration.

If data other than habitat quantity, quality, and location are available to measure the potential for exposure of assessment endpoints to stressors, using the alternative ranking scheme is more likely to be appropriate. Use of the alternative method under these circumstances would still characterize exposure, while avoiding the potential for underestimating risk in small subregions.

Information about the population dynamics of biological assessment endpoints can also help clarify which habitat ranking method is most suitable. If evidence exists that a population in a small patch of habitat in one subregion is indeed isolated from other populations in the study area, then the alternative ranking system remains the most appropriate. If organisms utilizing a small patch of habitat are simply a continuation of a population in adjacent subregions, the alternative method could potentially *overestimate* risk in subregions with small amounts of habitat, and the original habitat ranking scheme remains the most appropriate.

In the case of this RRM risk assessment, the CP subregion was appreciably smaller and therefore contained a smaller amount of habitat than the other subregions in the study area. This may account for the medium risk score for the CP subregion in the original RRM predictions. Using the alternative habitat ranking scheme increased the risk prediction for this subregion, indicating that the original RRM may have underestimated risk in the CP subregion. Additionally, limited data were available to characterize exposure to stressors in subregions; habitat quantity and quality remained the chief means for characterizing exposure. No data were available to evidence small isolated populations in subregions, additionally supporting the use of the original habitat ranking method. Applying the alternative ranking scheme as an exercise in uncertainty analysis revealed the possible underestimation of risk in the CP subregion in the original RRM.

The alternative habitat ranking scheme also demonstrated the broad applicability and flexibility of the RRM in performing regional-scale ecological risk assessments. Future RRM risk assessment should apply the habitat ranking method most appropriate for the study area. The availability of exposure and population data can help practitioners of the RRM determine which method to use. Furthermore, the question being asked should drive the selection of a ranking scheme. Risk assessments for endangered species are different than risk assessments for other endpoints, and the possibility of over- or underestimating risk should guide risk managers and assessors to choose a ranking method which is the most appropriate.

CONCLUSIONS

The RRM has been successful in multiple diverse settings, including marine ecosystems and inland watersheds (Landis and Wiegers 1997; Wiegers et al. 1998; Landis et al. 2000b; Walker et al. 2001; Obery and Landis 2002; Chen 2002; Moraes et al. 2002). These investigations, in combination with the current study, demonstrate the broad applicability and robustness of the RRM for use in regional risk assessment. Additionally, the Cherry Point RRM established the use of Monte Carlo techniques to better communicate uncertainty in an RRM regional risk assessment.

The regional ecological risk assessment for Cherry Point using the RRM characterized relative risk from anthropogenic sources in the region. The risk results suggested the major contributors of risk in the region are commercial and recreational vessel traffic, upland urban and agricultural landuse, and shoreline recreational activities. The biological endpoints most likely to be at risk are great blue heron and juvenile Dungeness crab. The majority of risk occurs in sandy intertidal, eelgrass, macroalgae, and soft bottom subtidal habitats. Spatially, the most risk occurs in the Lummi Bay, Drayton Harbor, and Cherry Point subregions. The risk predictions produced in this assessment are testable hypotheses on which to base future field and laboratory research in the region to reduce uncertainty and refine risk predictions in the region. The Cherry Point assessment characterized regional risk to multiple biological endpoints in the region and described the uncertainty associated with those risk predictions, providing valuable information about the spatial distribution of risk to local regional land managers unavailable in past assessments in the Cherry Point region.

REFERENCES

Alexander, R.R., Standon, R.J., Jr. and Dodd, J.R. 1993. Influence of sediment grain size on the burrowing of bivalves: correlation with distribution and stratigraphic persistence of selected neogene clams, *Palaios*, 8, 289–303.

Bellward, G.D., Norstom, R.J., Whitehead, P.E., et al. 1990. Comparison of polychlorinated dibenzodioxin levels with hepatic mixed-function oxidase induction in great blue herons, *J. Toxicol. Environ. Health*, 30, 33–52.

Brodeur, R.D. and Pearcy, W.G. 1992. Effects of environmental variability on trophic interactions and food web structure in a pelagic upwelling ecosystem, *Mar. Ecol. Progr. Ser.*, 84, 101–119.

Buchanan, D., Millemann, R.E., and Steward, N.E. 1970. Effects of the insecticide Sevin on various life stages of the Dungeness crab, *Cancer magister*, *J. Fish. Res. Board Can.*, 27, 93–104.

Burmaster, D.E. and Anderson, P.D. 1994. Principles of good practice for the use of Monte Carlo techniques in human health and ecological risk assessments, *Risk Anal.*, 14, 477–481.

Butler, R.W. 1995. The patient predator: foraging and population ecology of the great blue heron *Ardea herodius* in British Columbia. Occasional paper, number 86. Canadian Wildlife Service, Ottowa, Ontario.

Caldwell, R.S., Armstrong, D.A., Buchanan, D.V. et al. 1978. Toxicity of the fungicide Captan to the Dungeness crab, *Cancer magister*, *Mar. Biol.*, 48, 11–17.

Chapman, P.M., Dexter, R.N., Kocan, R.M., and Long, E.R. 1985. An Overview of Biological Effects Testing in Puget Sound, Washington: Methods, Results and Implications. *Aquatic Toxicology and Hazard Assessment: Seventh Symposium*, ASTM STP 854. Cardwell, R.D., Purdy, R., and Bahner, R.C. (Eds.). American Society for Testing and Materials, Philadelphia, PA, 344–363.

Chen, J. 2002. Regional Risk Assessment for the Squalicum Creek Watershed. Master of Science Thesis. Western Washington University, Bellingham, WA.

Collier, T.K., Stei, J.E., Sanborn, H.R. et al. 1992. A field study of the relationship between bioindicators of maternal contaminant exposure and egg and larval viability of English sole (*Parophrys vetulus*), *Mar. Environ. Res.*, 35, 171–175.

Cook, R.B., Suter, G.W., II, and Sain, E.R. 1999. Clinch River Series. Ecological risk assessment in a large river-reservoir: 1. introduction and background, *Environ. Toxicol. Chem.*, 18, 581–588.

Cormier, S.M., Smith, M., Norton, S., and Neiheisel, T. 2000. Assessing ecological risk in watersheds: a case study of problem formulation in the Big Darby Creek watershed, Ohio, USA, *Environ. Toxicol. Chem.*, 19, 1082–1096.

Ecology (Washington Department of Ecology). 2001. Accidental spills database for spills in Whatcom County 1995–2001. Unpublished data.

Ecology (Washington Department of Ecology). 1999. Sediment Quality in Puget Sound, Year 1— Northern Puget Sound, December 1999. Washington State Department of Ecology Environmental Assessment Program, Environmental Monitoring and Trends Section. Olympia, WA. Publication No. 99-347.

Ecology (Washington Department of Ecology). 1998. Final 1998 Section 303 (d) List – WRIA 1. Washington Department of Ecology, Olympia, WA.

Eissinger, A. 1998. ARCO Heron Project 1998 Annual Report. Prepared for ARCO Products Company Cherry Point Refinery. Nahkeeta Northwest Wildlife Services, Bow, WA.

Eissinger, A. 1996. Great blue herons of the Salish Sea: A Model Plan for the Conservation and Stewardship of Coastal Heron Colonies. Prepared for HeronLink, a joint venture between Trillium Corporation, ARCO Products Company and the Washington Department of Fish and Wildlife. Nahkeeta Northwest Wildlife Services, Bow, WA.

Elliot, J.E., Butler, R.H., Norstom, R.J., and Whitehead, R.E. 1989. Environmental contaminants and reproductive success of great blue heron *Ardea herodius* in British Columbia, *Environ. Pollut.*, 59, 91–114.

EVS Environmental Consultants. 1999. Cherry Point Screening Level Ecological Risk Assessment. Prepared for Washington Department of Natural Resources, Aquatic Resources Division, Olympia, WA

Gunderson, D.R., Armstrong, D.A., Shi, Y., and McConnaughey, R.A. 1990. Patterns of estuarine use by juvenile English sole (*Parophrys vetulus*) and Dungeness crab (*Cancer magister*), *Estuaries*, 13, 59–71.

Hart, L.E., Cheng, K.M., Whitehead, P.E. et al. 1991. Dioxin contamination and growth and development in great blue heron embryos, *J. Toxicol. Environ. Health*, 32, 331–344.

Healey, M.C. 1980. The eocology of juvenile salmon in Georgia Strait, British Columbia, in *Salmonid Ecosystems of the North Pacific*, McNeil, W.J. and Himsworth, D.C. (Eds.), Oregon State University Press, Corvallis, pp. 203–229.

Hirose, T. and Kawaguchi, K. 1998. Sediment size composition as an important factor in the selection of spawning site by the Japanese surf smelt *Hypomesus japonicus*, *Fish. Sci.*, 64, 995–996.

Holtby, L.B., Andersen, B.C., and Kadowaki, R.K. 1990. Importance of smolt size and early ocean growth to interannual variability in marine survival of coho salmon *Oncorhynchus kisutch. Can. J. Fish. Aquat. Sci.*, 47, 2181–2194.

Johnson, L.L., Casillas, E., Collier, T.K. et al. 1988. Contaminant effects on ovarian development in English sole (*Parophrys vetulus*) from Puget Sound, Washington, *Canad. J. Fish. Aquatic Sci.*, 45, 2133–2146.

Johnson, L.L., Casillas, E., Sol, S. et al. 1993. Contaminant effects on reproductive success in selected benthic fish, *Mar. Environ. Res.*, 35, 165–170.

Johnson, L.L., Landahl, J.T., Kubin, L.A. et al. 1998. Assessing the effects of anthropogenic stressors on Puget Sound flatfish populations, *J. Sea Res.*, 39, 125–137.

Johnson, L.L., Sol, S.Y., Ylitalo, G.M. et al. 1999. Reproductive injury in English sole (*Pleuronectes vetulus*) from Hylebos Waterway, Commencement Bay, Washington, *J. Aquatic Ecosys. Stress Recovery*, 6, 289–310.

Kelsall, J.P. 1989. *The Great Blue Heron of Point Roberts: History, Biology and Management*, self-published, Bellingham, WA.

Krygier, E.E. and Pearcy, W.G. 1986. The role of estuarine and offshore nursery areas for young English sole, *Parophrys vetulus* Girard, of Oregon, *Fishery Bull.*, 84, 119–132.

Kyte, M.A. 2001. Observations from a qualitative examination of the intertidal zone of the Cherry Point reach June 22–24, 2001. Golder Associates Inc., Bellevue, WA.

Kyte, M.A. 1994. The use by flatfish and Dungeness crabs of the Atlantic Richfield Company Cherry Point refinery vicinity. Pentec Environmental. Third annual report, project no. 138-002. Prepared for The Atlantic Richfield Company Cherry Point Refinery, Blaine, WA.

Kyte, M.A. 1993. The Use by Dungeness Crab and Flatfish of the BP Oil Company Ferndale Refinery Offshore Marine Environment. Pentec Environmental. Second annual report, 1992, on the post-dredging recovery study project no. 136-002. Submitted to BP Oil Company Ferndale Refinery, Ferndale, WA, USA.

Landis, W.G. and Wiegers, J.A. 1997. Design considerations and a suggested approach for regional and comparative ecological risk assessment, *Hum. Ecol. Risk Assess.*, 3, 287–297.

Landis, W.G., Luxon, M., and Bodensteiner, L.R. 2000b. Design of a relative risk model regional-scale risk assessment with confirmational sampling for the Willamette and McKenzie Rivers, Oregon, in *Environmental Toxicology and Risk Assessment: Recent Achievements in Environmental Fate and Transport: Ninth* Volume, Price, F.T., Brix, K.V., and Lane, N.K., Eds., ASTM STP 1381, 67–88. American Society for Testing and Materials, West Conshohocken, PA.

Landis, W.G., Markiewicz, A.J., Thomas, J., and Duncan, B. 2000a. Regional Risk Assessment for the Cherry Point Herring Stock. Western Washington University. Prepared for the Washington Department of Natural Resources, Aquatic Resources Division, Olympia, WA.

Laroche, W.A. and Holton, R.L. 1979. Occurrence of 0-age English sole, *Parophrys vetulus*, along the Oregon coast: an open coast nursery area? *Northwest Sci.*, 53, 94–96.

MacDonald, K., Simpson, D., Paulson, B. et al. 1994. Shoreline armoring effects on physical coastal processes in Puget Sound, Washington. Coastal Erosion Management Studies, Volume 5. Washington Department of Natural Resources, Shoreland and Water Resources Program, Report 94-78.

Malins, D.C., McCain, B.B., Brown, D.W. et al. 1985. Chemical pollutants in sediments and diseases of bottom-dwelling fish in Puget Sound, Washington, *Environ. Sci. Technol.*, 18, 705–713.

Markiewicz, A.J., Hart Hayes, E., and Landis, W.G. 2001. Cherry Point Herring Regional Risk Assessment Phase II. Western Washington University. Prepared for Washington Department of Natural Resources, Aquatic Resources Division, Olympia, WA.

McMillan, R.O. 1991. Abundance, Settlement, Growth and Habitat Use by Juvenile Dungeness Crab, *Cancer magister*, in Inland Waters of Northern Puget Sound, Washington. Master of Science Thesis. University of Washington, Seattle.

Miller, B.S., Simenstad, C.A., Moulton, L.L. et al. 1977. Puget Sound baseline program nearshore fish survey: final report, July 1974–June 1977. Baseline Study Report no. 10. Washington State Department of Ecology, Lacey, WA.

Moraes, R., Landis, W.G., and Molander, S. 2002. Regional risk assessment of a Brazilian rain forest reserve, *Hum. Ecol. Risk Assess.*, 8, 1779–1804.

Myers, M.S., Johnson, L.L., Olson, P.O. et al. 1998. Toxicopathic hepatic lesions as biomarkers of chemical contaminant exposure and effects in marine bottomfish species from the Northeast and Pacific coasts, USA, *Mar. Pollut. Bull.*, 37, 92–113.

NOS (National Ocean Service). 2001. National Geophysical Data Center GEODAS NOS Hydrographic Surveys [computer file], Vol. 1, Version 4.1.

NSEA (Nooksack Salmon Enhancement Association). 2000. 2000 Terrell Creek water quality and smolt trap data. Unpublished data.

Obery, A.M. and Landis, W.G. 2002. A regional multiple stressor risk assessment of the Codorus Creek watershed applying the relative risk model, *Hum. Ecol. Risk Assess.*, 8, 405–428.

Pauley, G.B., Armstrong, D.A., and Heun, T.W. 1986. Species profiles: life histories and environmental requirements of coastal fishes and invertebrates (Pacific Northwest) — Dungeness crab. U.S. Fish and Wildlife Service Report 82(11.63). U.S. Army Corp of Engineers, TR EL-82-4.

Pearson, W.H., Sugarman, P.C., Woodruff, D.L., and Blaylock, J.W. 1980. Detection of petroleum hydrocarbons by the Dungeness crab, *Cancer magister*, *Fish. Bull.*, 78, 821–826.

Pentilla, D.E. 1997. Investigations of Intertidal Spawning Habitats of Surf Smelt and Pacific Sand Lance in Puget Sound, Washington. Proceedings of the International Symposium on the Role of Forage Fishes in Marine Ecosystems. 13–16 November 1996. Anchorage, AK.

Rhodes, L. and Casillas, E. 1985. Interactive effects of cadmium, polychlorinated biphenyls, and fuel oil on experimentally exposed English sole (*Parophrys vetulus*). *Can. J. Fish. Aquatic Sci.*, 42, 1870–1880.

Shi, Y., Gunderson, D.R., and Sulivan, P.J. 1997. Growth and survival of 0+ English sole, *Pleuronectes vetulus*, in estuaries and adjacent nearshore waters of Washington, *Fish. Bull.*, 95. 161–173.

Simenstad, C.A., Tanner, C.D., Thom, R.M., and Conquest, L.L. 1991. Estuarine habitat assessment protocol. USEPA Region 10, Office of Puget Sound. Seattle, WA. EPA 910/9-91-037.

Stein, J.E., Hom, T., Sanborn, H.R., and Varanasi, U. 1991. Effects of exposure to a contaminated-sediment extract on the metabolism and disposition of 17β-estradiol in English sole (*Parophrys vetulus*), *Compar. Biochem. Physiol.*, 99C, 231–240.

Suter, G.W. (Ed.). 1993. *Ecological Risk Assessment*, Lewis Publishers, Chelsea, MI, p. 365.

Thom, R.M. and Shreffler, D.K. 1994. Shoreline Armoring Effects on Coastal Ecology and Biological Resources in Puget Sound, Washington. Coastal Erosion Management Studies, Volume 7. Washington Department of Ecology, Shoreland and Water Resources Program, Report 94-80.

Toole, C.L., Barnhart, R.A., and Onuf, C.P. 1987. Habitat Suitability Index Models: Juvenile English Sole. U.S. Fish and Wildlife Service Biological Report 82(10.133).

USEPA (U.S. Environmental Protection Agency). 2001. Better Assessment Science Integrating Point and Nonpoint Sources [CD ROM]. Office of Water, Washington, D.C., EPA-823-C-01-010.

USEPA (U.S. Environmental Protection Agency). 1998. Guidelines for Ecological Risk Assessment. Washington, D.C., EPA/360/R-95/002F.

USGS (U.S. Geological Survey). 1987. Digital Line Graph of Bellingham, Washington [computer file]. Using: USGS Bellingham, Washington (map) 1:100,000. 30-minute series. Reston, VA: USGS, 1975.

Walker, R., Landis, W.G., and Brown, P. 2001. Developing a regional ecological risk assessment: a case study of a Tasmanian agricultural catchment, *Hum. Ecol. Risk. Assess.*, 7, 417–439.

Warren-Hicks, M.J. and Moore, D.R.J. 1998. Uncertainty analysis in ecological risk assessment. Proceedings from the Pellston workshop on uncertainty in ecological risk assessment, 23–28 August 1995, Pellston, Michigan. Society of Environmental Toxicology and Chemistry (SETAC). Pensacola, FL.

WDFW (Washington Department of Fish and Wildlife). 2002a. WDFW surf smelt information page. Available at http://www.wa.gov/wdfw/fish/forage/smelt.htm#shabitat.html.

WDFW (Washington Department of Fish and Wildlife). 2002b. Birch Bay State Park Shellfishing Guidelines. Available at http://www.wa.gov/wdfw/fish/shelfish/beachreg/b200060.html.

WDFW (Washington Department of Fish and Wildlife). 2002c. Priority Habitats and Species List. Available at http://www.wa.gov/wdfw/hab/phslist.html.

WDFW (Washington Department of Fish and Wildlife). 2001a. Aerial observations of recreational clam and pacific oyster harvest on Puget Sound public beaches from January 1 through December 31, 2001, Washington Department of Fish and Wildlife, Olympia, WA.

WDFW (Washington Department of Fish and Wildlife). 2001b. 1999–2001 Ballast Water Release Records. Unpublished data.

WDFW (Washington Department of Fish and Wildlife). 2001c. 2001 Surf Smelt Spawning Areas [computer file]. Washington Department of Fish and Wildlife, Olympia, WA.

WDFW (Washington Department of Fish and Wildlife). 1999. Great Blue Heron Densities from the Puget Sound Ambient Monitoring Program Summer Aerial Counts 1992–1999. Washington Department of Fish and Wildlife, Olympia, WA.

WDFW (Washington Department of Fish and Wildlife). 1998. Littleneck Clam Population Survey at Birch Bay State Park, 1998. Washington Department of Fish and Wildlife, Olympia, WA.

WDNR (Washington Department of Natural Resources). 2001a. DNR Aquatic Mission. Available at http://www.wa.gov/dnr/htdocs/aqr/html.

WDNR (Washington Department of Natural Resources). 2001b. Cherry Point Reserve Boundaries [computer file]. Washington Department of Natural Resources, Olympia, WA.

WDNR (Washington Department of Natural Resources). 1997a. Shorezone Inventory. Nearshore Habitat Program [computer file]. Washington Department of Natural Resources, Olympia, WA.

WDNR (Washington Department of Natural Resources). 1997b. Puget Sound Intertidal Habitat Inventory 1995 [CD-ROM]. Washington State Department of Natural Resources Nearshore Habitat Program, Aquatic Resources Division, Olympia, WA.

Whatcom County Assessor. 2000. 2000 tax parcels [computer file]. Whatcom County Assessor, Bellingham, WA.

Whatcom County Planning and Development Services. 2001. Water Resource Inventory Area 1 Watershed boundaries [computer file]. Whatcom County Planning and Development Services, Bellingham, WA.

Whatcom County Planning and Development Services. 1998. Buffered Wetlands [computer file]. Using: U.S. Fish and Wildlife Services National Wetlands Inventory. Whatcom County Planning and Development Services, Bellingham, WA.

Whatcom County PUD (Public Utility District). 2000. Updated land use parcels for Lummi Nation Indian Reservation [computer file]. Whatcom County Public Utility District, Lynden, Washington.

Wiegers, J.K., Feder, H.M., Mortensen, L.S. et al. 1998. A regional multiple-stressor rank-based ecological risk assessment for the fjord of Port Valdez, Alaska, *Hum. Ecol. Risk Assess.*, 4, 1125–1173.

Index